Springer Tracts in Advanced Robotics

Volume 160

Frank Chongwoo Park, Mechanical Engineering Department, Seoul National University, Seoul, Korea (Republic of)

S. E. Salcudean, The University of British Columbia, Vancouver, BC, Canada

Roland Siegwart, LEE J205, ETH Zürich, Institute of Robotics & Autonomous Systems Lab, Zürich, Switzerland

Gaurav S. Sukhatme, Department of Computer Science, University of Southern California, Los Angeles, CA, USA

The Springer Tracts in Advanced Robotics (STAR) publish new developments and advances in the fields of robotics research, rapidly and informally but with a high quality. The intent is to cover all the technical contents, applications, and multidisciplinary aspects of robotics, embedded in the fields of Mechanical Engineering, Computer Science, Electrical Engineering, Mechatronics, Control, and Life Sciences, as well as the methodologies behind them. Within the scope of the series are monographs, lecture notes, selected contributions from specialized conferences and workshops, as well as selected PhD theses.

Special offer: For all clients with a print standing order we offer free access to the electronic volumes of the Series published in the current year.

Indexed by SCOPUS, DBLP, EI Compendex, zbMATH, SCImago.

All books published in the series are submitted for consideration in Web of Science.

Chenguang Yang · Zhenyu Lu · Ning Wang

Advanced Teleoperation and Robot Learning for Dexterous Manipulation

Springer

Chenguang Yang
Department of Computer Science
University of Liverpool
Liverpool, UK

Zhenyu Lu
School of Automation Science
and Engineering
South China University of Technology
Guangzhou, China

Ning Wang
Department of Computing
Sheffield Hallam University
Sheffield, UK

ISSN 1610-7438 ISSN 1610-742X (electronic)
Springer Tracts in Advanced Robotics
ISBN 978-3-031-78500-9 ISBN 978-3-031-78501-6 (eBook)
https://doi.org/10.1007/978-3-031-78501-6

This Springer imprint is published by the registered company Springer Nature Switzerland AG
The registered company address is: Gewerbestrasse 11, 6330 Cham, Switzerland

Series Editor's Foreword

At the dawn of the century's third decade, robotics is reaching an elevated level of maturity and continues to benefit from the advances and innovations in its enabling technologies. These all are contributing to an unprecedented effort to bringing robots to human environment in hospitals and homes, factories, and schools, in the field for robots fighting fires, making goods and products, picking fruits, watering the farmland, and saving time and lives. Robots today hold the promise for making a considerable impact in a wide range of real-world applications from industrial manufacturing to healthcare, transportation, and exploration of the deep space and sea. Tomorrow, robots will become pervasive and touch upon many aspects of modern life.

The *Springer Tracts in Advanced Robotics (STAR)* is devoted to bringing to the research community the latest advances in the robotics field on the basis of their significance and quality. Through a wide and timely dissemination of critical research developments in robotics, our objective with this series is to promote more exchanges and collaborations among the researchers in the community and contribute to further advancements in this rapidly growing field.

The monograph by Chenguang Yang, Zhenyu Lu and Ning Wang is focused on telerobotics and robot learning, and its contents are organized in eight chapters. Several methods for dexterous manipulation based on a single-leader—dual-follower (SLDF) approach for teleoperation are proposed using autonomous skill learning techniques, including bio-inspired strategies. The main challenge is to achieve the integration of autonomous learning algorithms within the control framework.

A nice blend of original theoretical and practical results in the field of dexterous robotic manipulation, this volume is a great addition to the STAR series.

Naples, Italy
July 2024

Bruno Siciliano
STAR Editor

Preface

During the last decades, robots have been expected to become increasingly intelligent to deal with different types of tasks. More specifically, humans expect that robots can acquire adaptive manipulation skills effectively and efficiently. In robotic manipulation, dexterous manipulation is an area in which multiple manipulators, or fingers, cooperate to grasp and manipulate objects. A distinguishing characteristic of dexterous manipulation is that it is object-centered, which means the problem is formulated in terms of the object being manipulated, and the forces that need to be applied to it. Robotic teleoperation (also called telerobotic) is an efficient way to realize natural dexterous manipulation. In most of the existing teleoperation methods, leader-follower mode is the basic framework, where the operator manipulates a leader robot to telemanipulate or send commands to the remote follower robot(s).Within this mode, the single-leader-dual-follower (SLDF) teleoperation system allows two follower robots to cooperate with each other under the control instructions from the leader side. To achieve dexterous manipulation, the SLDF system not only needs to track the trajectory planned by the human operator, but also to intelligently adjust its exerting force according to the properties of the interactive objects. This is also the area we are mainly focused on. In this book, we will introduce several methods to implement dexterous manipulation based on the SLDF system.

Robotic skill learning and control strategy are also important for dexterous manipulation. The human neural motor system has the intelligence to learn new skills and control strategies, and then to generalize this knowledge naturally. Such an intelligent process will involve the reuse and migration of knowledge, but it is not easy for a robot to demonstrate. Inspired by the neural motor system, skill learning and control strategy frameworks based on integrated automation ideas have the effect of knowledge reuse and autonomous adaptation to new environments. In addition, as in other fields, bio-inspired is also applicable to realize autonomous skill learning and control to the robots, and the bio-inspired strategies will give the robots more anthropomorphic intelligence. In this book, we will introduce our latest research on autonomous learning and control frameworks. We will show that with the help of these approaches, the robot can autonomously learn and generalize the skills, together with enhancing its control performance during the task generalization and execution, showing the potential of robots in autonomy and anthropomorphic.

This book will be primarily focused on the research in dexterous manipulation.

The structure of this book is summarized below:

In Part I, we will conduct our work around the SLDF system in robotic teleoperation. Specifically, we will explore the challenges of applying the SLDF system to the grasping and holding tasks of nonrigid objects and provide some efficient techniques to solve these problems. With the help of these techniques, the assistive follower robot (AFR) in SLDF will be ensured to closely cooperate with the direct-teleoperated follower robot (DFR), while maintaining appropriate contact force with the object. Chapter 1 will mainly focus on designing a force controller for the AFR based on an autoregressive model and impedance control, which enables the followers to hold a rigid or a deformable object with a desired contact force. Chapter 2 will solve the problems using an adaptive movement estimation approach will be used to enable the AFR to infer its partner's movement according to the contact force, and the impedance controller will be derived such that the desired contact force can be achieved.

In Part II, we will present several works on integrated autonomous learning and control frameworks. Radial basis function neural network (RBFNN) has a good performance of function fitting and is frequently applied to modeling the unknown item of the robot dynamics system. Traditional RBFNN often needs to retrain the parameters after facing new inputs from the new environment, requiring a certain computing cost and time cost. Thus, we will put our attention on the improvement of RBFNN so that it can reuse the learned parameters and increment the network at a small cost. In Chap. 3, we will begin with an improved incremental RBFNN framework to estimate the manipulator dynamic parameters, and then with the help of the Lyapunov function, we will show a stable control strategy that could converge within a fixed time. Finally, a safety region convolutional neural network (CNN) method adapted to the remote center of motion method will be proposed to achieve manipulation in the small opening workspace without collision. In Chap. 4, we will propose a framework of broad learning based on novel adaptive neural control, such that in the presence of dynamic disturbance, robots can learn a set of basic skills and then generalize these skills to the neighboring movements naturally as humans, by incorporating the deterministic learning with broad learning system and adaptive neural control. Besides, to complete the better learning of the skills, in Chap. 5, we will display incremental robot skill learning and generalization framework including an incremental dynamical movement primitives (IDMP) for robot trajectory learning and an adaptive neural network control method, which are incrementally updated to enable robots to adapt to new cases. This new control strategy is incrementally updated and can accumulate and reuse the learned knowledge to improve learning efficiency.

In Part III, several bio-inspired autonomous learning for dexterous manipulations will be introduced. For detail, in Chap. 6, we will propose a novel DMP framework facing the constrained conditions for robotic skill generalization. Barrier Lyapunov functions (BLFs) will be deduced to compensate for tracking errors between the real and desired trajectories with constraints in DMP's acceleration term. In Chap. 7, a broad learning system (BLS) is proposed for extracting both ordinary skills and instant reactive skills from demonstrations, the latter of which is usually generated to avoid a sudden danger (e.g., touching a hot cup). In Chap. 8, a novel control strategy based on a broad fuzzy neural network (BFNN) which is subjected to contact with an unknown environment will be proposed. The BFNN can use the advantages of a BLS, which can dynamically adjust the number of neurons to approach the unknown robot dynamic model. For external unknown interference, we develop an adaptive impedance learning method that can iterate the optimal feedback gain in a limited time.

Liverpool, UK — Chenguang Yang
Guangzhou, China — Zhenyu Lu
Sheffield, UK — Ning Wang

Acknowledgements Many insightful suggestions and comments from our colleagues, friends, and coworkers have helped us to improve this book. Here, we would like to express our sincere gratitude to them. We would like to express our appreciation to those who have contributed to the collaborative research reported in this monograph. We would like to thank Weiyong Wang, Yubo Dong, Wenjin Xu, and Lingzi Xie who supported for the organization of the materials.

Contents

Part I
Single-Leader-Dual-Follower Teleoperation for Cooperative Manipulation

Chapter 1
Motion Regulation Solutions for Holding and Moving Objects in Single-Leader-Dual-Follower Teleoperation

1.1 Introduction

Robot teleoperation has been widely investigated in various applications [1], including healthcare [2] and aerospace [3], etc. Most of the existing teleoperation methods are based on a leader-follower mode, where the operator manipulates a leader to tele-manipulate or send commands to the remote follower robot(s) [1]. Within this mode, the single-leader-dual-follower (SLDF) teleoperation system, as shown in Fig. 1.1, has drawn numerous researchers' attention recently because it has a unique advantage of alleviating the operator's workload while conducting almost the same amount of work [4], compared to the basic single-leader-single-follower system [5], the multiple-leader-multiple-follower system [6–8] and the multiple-leader-single-follower system [9, 10]. In an SLDF teleoperation system, the two follower robots are usually divided into one direct-teleoperated follower robot (DFR) and an assistive follower robot (AFR). Researches in this area mostly focus on enabling AFR to assist in completing a task autonomously, without any instruction from the operator. For example, a semi-autonomous SLDF framework was proposed in [11] to conduct suturing tasks in the minimally invasive surgery (MIS); an automated relay SLDF technique was devised in MIS [12]; a multi-layer controller was developed in [4] to generate target poses and forces for the followers to perform manipulation of a hard-contact object. Despite the great achievement in the literature, there is still room to improve the current methods, for example, in [4], the desired movement is obtained according to tasks instead of the operator's motion, which lacks of versatility in response to the operator's motion intent.

Besides teleoperation, the SLDF is also found in the field of multiple robot systems (MRS) since there are two followers in the system. Note that in this chapter we discuss manipulators' MRS [13], other than mobile robots' MRS. In this field, there are many papers on using MRS to transport a rigid object, but few works about holding and moving a non-rigid volumetric object [14]. There is a challenge for the SLDF system to hold and move a volumetric-type deformable object [15]: to generate a reference

C. Yang et al., *Advanced Teleoperation and Robot Learning for Dexterous Manipulation*, Springer Tracts in Advanced Robotics 160,
https://doi.org/10.1007/978-3-031-78501-6_1

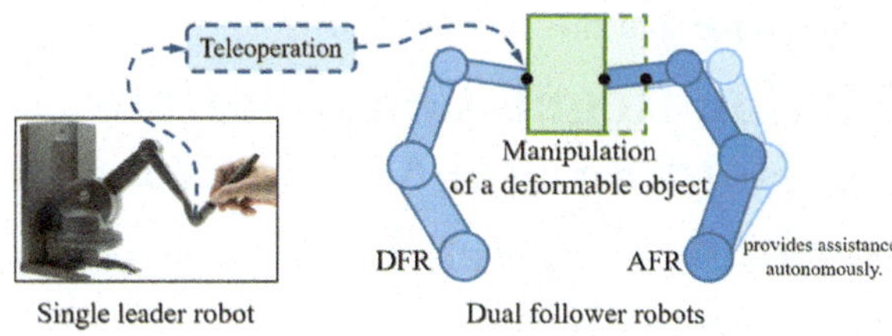

Fig. 1.1 An SLDF teleoperation system where one leader robot is employed to teleoperate two follower robots to hold and move an object

trajectory for AFR according to the operator's random movement, while the desired holding force is required to be maintained.

To improve the ability of motion regulation of a robot, the artificial potential field method (APFM) has been verified to be a practical means. Its use is found to cover a large number of application scenarios, which are mostly about mobile robots' or unmanned aerial vehicles' (UAVs') [16] path planning. There are a few works about APFM for manipulators' motion regulation [17, 18]. Researchers have made great efforts to enhance the performance of an APFM, e.g., the reinforcement learning technique is integrated in [19] to deal with moving obstacles. However, it can be more challenging if the regulation of the manipulator's target poses takes human factors (discussed later) into consideration, which is critical in SLDF teleoperation.

The problem of the internal force regulation in an object-holding task for MRS is different from a pick-and-place grasping task for only a single robot [20], because it requires planning coordination trajectories and control to execute the collaborating holding movements for at least two manipulators. Researchers have put forward methods targeting certain aspects. For example, in [21], the proposed task-space regulators maintain a desired internal force of the rigid object held by two collaborative manipulators. These methods share features in common: i) the manipulated objects are rigid and ii) the holding pose is horizontal, i.e., the gravity does not affect the contact force. To extend the MRS's ability to hold and move especially volumetric deformable objects[15], different physical models of deformable objects are introduced in [22]. Besides other kinds of deformable objects, e.g., in [23], S. Zimmermann et al. generated optimal trajectories for one or both robot arms for dynamic manipulation, such as swinging a soft rod as a whip, a pendulum as a wrecking ball to knock off other objects, laying soft sheets onto a table, which are model-based and the whole trajectories are calculated ahead of execution, there have been some works concentrating on the manipulation of these volumetric deformable objects. A generalized learning algorithm for dual robots to estimate the object's deformation profiles was developed in [24], while Z. Lu et al. mainly focused on designing a position predictor and a force predictor under time-delay and F/T sensor-less condition for the dual-arm tele-robot to hold and move a yoga ball in one axis [25]. These works are likewise limited to the condition that the object is held in a pre-defined and fixed manipulation pose. However, in a hold-and-move SLDF teleoperation scenario, the contact force varies because of influences of the object's mass, varying internal force and the operator's random pick-up poses, holding poses and movements (human factors), etc. Those predefined-trajectory methods are doomed to fail as the manipulating pose is not fixed and unpredictable.

To propose a complete set of solutions for an SLDF system to hold and move an object, the contributions include:

- We derive a series of techniques to regulate two followers based on a single operator's motion. First, the re-"fixing" pose transformation enables precise/flexible and ergonomic teleoperation, which also reduces the tedious whole-task-space matching work. Second, we provide a stability-proved efficient technique to obtain the holding orientation for AFR via DFR's quaternions. Third, we design an adjustable APFM, which takes the operator's random motions into consideration to update weights. These adjustable weights endow AFR with intelligence to judge whether to assist holding or prepare to assist holding, despite random holding poses specified by the operator.
- We extend our previous publication [26] to the SLDF teleoperation system for the first time, where an autoregressive model and the impedance model are combined to generate AFR's reference trajectory such that AFR can actively move to the right position to track the desired contact force, regardless of the decoupled DFR's random movement and the object's unknown parameters and time-varying internal forces. The reference trajectory can be directly applied without the need to wait for parameter estimation to converge.

1.2 Problem Formulation

1.2.1 System Description

An SLDF teleoperation system generally consists of three robots, as depicted in Fig. 1.1. One serves as the leader and the other two as followers. The leader is manipulated by the operator and two followers are teleoperated according to the single leader. Two followers usually play different roles when carrying out a task, where one (DFR) is directly teleoperated by the leader and the other (AFR) provides assistance to DFR.

A robot teleoperation system usually applies a mapping between the task spaces (1.1) or the joint spaces (1.2) of the leader and the follower:

$$X_{follower} = f(X_{leader}), \tag{1.1}$$

$$\theta_{follower} = g(\theta_{leader}), \tag{1.2}$$

where $X_*, * = leader, follower$ stands for the end-effector (EE) pose of the follower or the leader, θ_* are their joint positions and $f(\cdot)$ and $g(\cdot)$ are the mappings.

Compared to joint space transformation (1.2) which may encounter problems such as mismatching degrees of freedom (DOF), task space transformation in (1.1) is preferable.

When task space movement commands are received, target joint positions can be computed via inverse-kinematics methods, e.g., the closed-loop inverse kinematics [27]:

$$\dot{\theta}_d = J^{\dagger}(\dot{X}_d + k_E(X_c - X_d)), \tag{1.3}$$

where θ_d is the desired joint position leading to the desired EE position X_d, X_c is EE's current position, k_E is a user-defined gain and $J^{\dagger}$ is the inverse/pseudo-inverse of the robot's Jacobian matrix J.

Note that all poses/positions, etc. mentioned in the following are about EE, but "EE" is left out for brevity.

1.2.2 *Problem Statement*

In the SLDF teleoperation for objects hold-and-move tasks, (i) DFR is expected to be teleoperated ergonomically, i.e., with easeful gesture/posture in the operator's hands/arms and without causing any pain or discomforts to any arthrosis or muscle; (ii) AFR should be able to autonomously regulate its motion both before and during holding the object; and (iii) their collaboration needs to maintain a certain holding force. These expectations can be achieved if the following three problems are addressed (Fig. 1.2).

Firstly, awkward postures, e.g., upper arm abduction, contribute to musculoskeletal disorders [28, 29] and many results reveal that it is important to re-design the tasks involving static or repetitive forearm twisting, based on ergonomic analysis [30]. In a typical screw driving scenario (with repetitive pronation/supination motions) [29], the method of mapping leader's entire task space to the follower's can lead to problems, e.g., as shown in Fig. 1.3, where the awkward gesture (the right palm twists

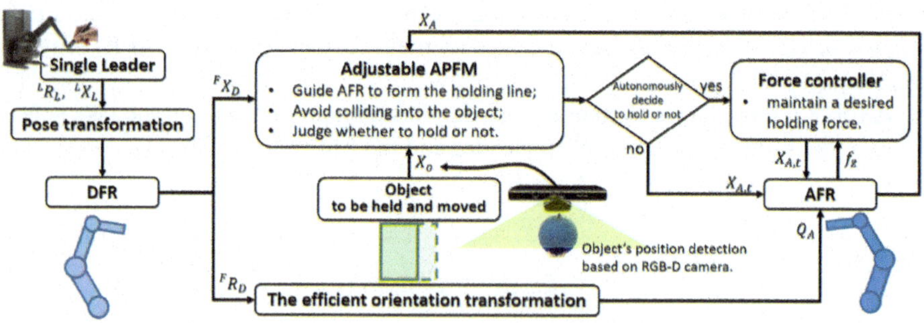

Fig. 1.2 The flowchart of the whole structure of the proposed methods

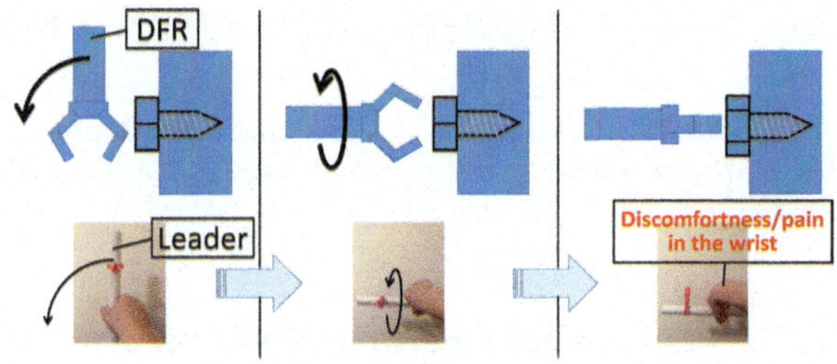

Fig. 1.3 A scenario where the operator tightens a nut, which requires him/her to keep an uncomfortable pose with the right palm twisting toward the outsides of the body

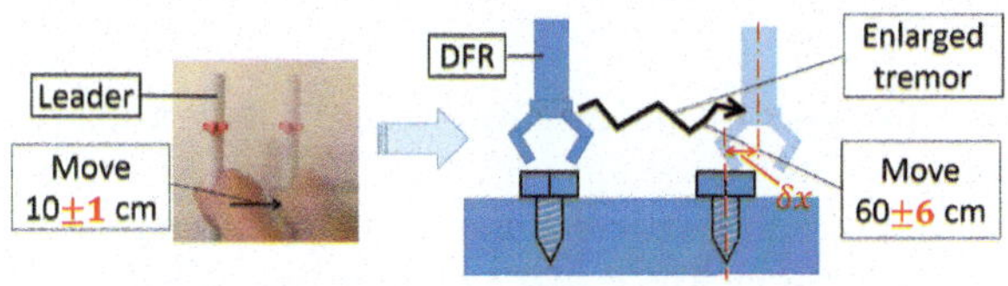

Fig. 1.4 A scenario where the tremor and the inaccuracy of the operator's movement are enlarged, affecting a precise operation

toward outsides the body) causes discomfort to the operator or even pain to his/her wrist/forearm if keeping using the same gesture for long. Also seen in Fig. 1.4, it is hard for the operator to teleoperate DFR to precisely move along a short trajectory since industrial manipulators (followers) usually have a much larger workspace than a desktop joystick (the leader). In this situation, the enlarged movement also enlarges the operator's tremors and manual motion errors. Therefore, a new design to facilitate an ergonomic teleoperation to achieve precise/flexible manipulation is in demand.

Secondly, to alleviate the operator's workload, AFR should actively regulate its motion to assist DFR teleoperated by the operator. However, when the object moves irregularly and as human is in the loop, it is challenging to achieve a proper pose for AFR to fit the following requirements: i) tracking and avoiding collision with the object; ii) adjusting a pose ready to collaborate with DFR; iii) judging whether to execute holding or keep waiting.

Finally, when the contact force on the deformable object is required to be maintained within a small range, it can be challenging to generate a collaborative movement for AFR when the object's shape changes, its mass and stiffness are unknown and DFR's (the operator's) movement is not pre-defined and subjected to uncertainties.

Theorem 1.1 *Before the following discussion, we assume that all followers perfectly track the commanded poses that they receive, with no time delay. And the teleoperation precision discussed is the precision under the circumstance that the problems caused by time delays are assumed to be already tackled.*

1.3 Motion Regulation Before Object Holding

This section corresponds to the first contribution, which contains the motion regulation methods for both the orientation and the position of DFR and AFR.

1.3.1 Direct-Teleoperated Follower Robot (DFR)'s Orientation

For the sake of universality, we use rotation matrices and quaternions to discuss orientations.

In Fig. 1.5, the rotation matrix ${}^F_L R$ between bases of the leader and DFR is known via calibration. The leader's rotation matrix ${}^L R_{L,t}$ can be read in real time. ${}^L R_{L,t_0}$ is an initial pose set at any time t_0 ($t > t_0$), corresponding to DFR's orientation ${}^F R_{D,t_0}$ at the same time. To derive DFR's orientation when the leader rotates from t_0 to t, we let DFR's rotation be consistent with the leader's after t_0 by "fixing" them together at t_0. In other words, their relative pose R_{fixed,t_0} should stay constant from t_0 on. Note that it is easy to repeatedly "fix" a pose by a press on a keyboard or a click on a mouse.

For the inner closed transforming chain (ABCD and AD in Fig. 1.5), we have

$$ {}^F_L R \, {}^L R_{L,t_0} R_{\text{fixed},t_0} = {}^F R_{D,t_0}. \tag{1.4} $$

Then, the fixed relative rotation is obtained

$$ R_{\text{fixed},t_0} = \left({}^F_L R^L R_{L,t_0} \right)^{-1} {}^F R_{D,t_0}. \tag{1.5} $$

For the outer closed transforming chain (ABEF and AF in Fig. 1.5), letting $R_{\text{fixed},t} = R_{\text{fixed},t_0}$. Then, the target DFR's orientation at any time t is

$$ {}^F R_{D,t} = {}^F_L R \, {}^L R_{L,t} R_{\text{fixed},t_0}. \tag{1.6} $$

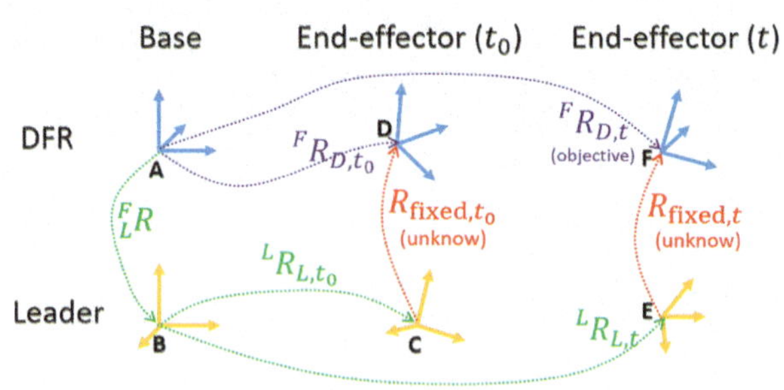

Fig. 1.5 Orientation transformation from the leader to DFR

By repeatedly re-setting the initial orientation at t_0, t_1, ..., the operator can reach any target DFR's orientation by using any desired pose to manipulate the joystick.

1.3.2 Direct-Teleoperated Follower Robot (DFR)'s Position

Followers share the same base frame. Then,

$$ {}^{F}X_D = s\,{}^{F}_{L}R\,{}^{L}X_L \tag{1.7} $$

where ${}^{F}X_D$ and ${}^{L}X_L$ are respectively the positions of DFR and the leader and s is a scaling coefficient.

Similarly, the relative displacements are:

$$ {}^{L}\Delta X_L = {}^{L}X_{L,t} - {}^{L}X_{L,t_0}, \tag{1.8} $$

$$ {}^{F}\Delta X_D = {}^{F}X_{D,t} - {}^{F}X_{D,t_0}, \tag{1.9} $$

where ${}^{F}X_{D,t}$ and ${}^{F}X_{D,t_0}$ are DFR's target position at t and the initial position set by the operator at t_0, respectively. Premultiplying both sides of (1.8) by $s\,{}^{F}_{L}R$, we have

$$ \begin{aligned} s\,{}^{F}_{L}R\,{}^{L}\Delta X_L &= s\,{}^{F}_{L}R({}^{L}X_{L,t} - {}^{L}X_{L,t_0}) \\ &= {}^{F}X_{D,t} - {}^{F}X_{D,t_0} = {}^{F}\Delta X_D. \end{aligned} \tag{1.10} $$

Then, we derive the target position of DFR referring to the user-defined initial position via

$$ {}^{F}X_{D,t} = {}^{F}X_{D,t_0} + {}^{F}\Delta X_D = {}^{F}X_{D,t_0} + s\,{}^{F}_{L}R\,{}^{L}\Delta X_D. \tag{1.11} $$

By repeatedly setting new but different initial positions, we can reach any desired position in DFR's task space. In practice, the setting of the scaling factor s can be achieved easily, e.g., via pressing the "+" or "−" button on a keyboard or by adaptively self-adjusting based on the moving speed of the leader, such that we can enlarge/shrink the relative displacement and thus achieve flexible/precise teleoperation.

1.3.3 Assistive Follower Robot (AFR)'s Orientation

The basic transformation from rotation matrices to unit quaternions is

$$ \begin{aligned} q_0 &= \frac{1}{2}\sqrt{1 + r_{11} + r_{22} + r_{33}}, \\ q_1 &= \frac{r_{32} - r_{23}}{4q_0}, q_2 = \frac{r_{13} - r_{31}}{4q_0}, q_3 = \frac{r_{21} - r_{12}}{4q_0}, \end{aligned} \tag{1.12} $$

where q_0, q_1, q_2, q_3 are elements of a quaternion Q and r_{ij} $(i, j = 1, 2, 3)$ is the ith row and jth column element of a rotation matrix. Detailed discussions on situations where $1 + r_{11} + r_{22} + r_{33} < 0$ can be found in literature, e.g., [31].

When DFR's orientation (in quaternion) Q_D is obtained from ${}^F R_{D,t}$ through (1.12), we acquire the desired AFR's quaternion Q_A by employing the computationally efficient algorithm in Lemma 1.1.

Lemma 1.1 *If the quaternion of DFR is*

$$Q_D = [q_{1,0}, q_{1,1}, q_{1,2}, q_{1,3}]^T, \tag{1.13}$$

AFR's quaternion for a common humanoid holding manner, with AFR's orientation pointing to DFR's end-effector and without generating unnecessary torques or twists acting on the object if both AFR and DFR are aligned on one line, is

$$Q_A = [q_{2,0}, q_{2,1}, q_{2,2}, q_{2,3}]^T = [-q_{1,1}, q_{1,0}, q_{1,3}, -q_{1,2}]^T. \tag{1.14}$$

And the proof of Lemma 1.1 is as follows:

For two robots to perform holding pose in a unified coordinate system, taking Fig. 1.6 as an example, they firstly need to align their EEs into a face-to-face pose, which results in rotating DFR and AFR about x axis $-90°$ and $+90°$ respectively. Then, to keep the object held properly at all time, AFR's and DFR's rotations should always stay in a relation described as

$$\theta_{x,A} = \theta_{x,D} + 180°, \theta_{y,A} = \theta_{y,D}, \theta_{z,A} = \theta_{z,D}, \tag{1.15}$$

where $\theta_{*,A}$ and $\theta_{*,D}$ are angles of AFR and DFR rotating about fixed $*$ axis respectively. As quaternions share the same calculation from Euler angles, substituting $\theta_{*,A}$ with $\theta_{*,D}$ in

Fig. 1.6 The relation between the initial pose and the holding pose of DFR and AFR referring to a unified coordinate system

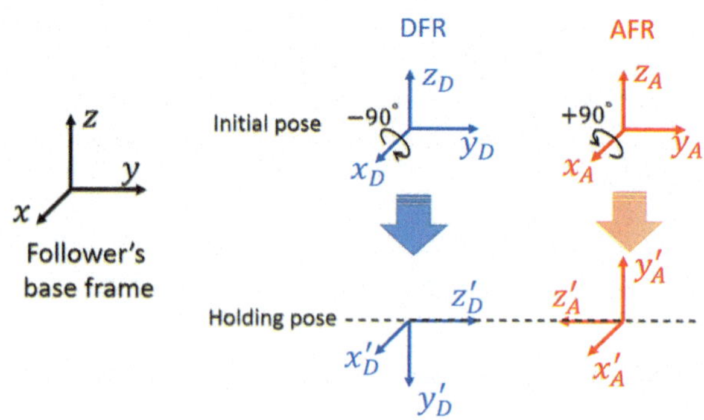

$$Q_A = \begin{bmatrix} q_{2,0} \\ q_{2,1} \\ q_{2,2} \\ q_{2,3} \end{bmatrix} = \begin{bmatrix} c\frac{\theta_{x,A}}{2}c\frac{\theta_{y,A}}{2}c\frac{\theta_{z,A}}{2} + s\frac{\theta_{x,A}}{2}s\frac{\theta_{y,A}}{2}s\frac{\theta_{z,A}}{2} \\ s\frac{\theta_{x,A}}{2}c\frac{\theta_{y,A}}{2}c\frac{\theta_{z,A}}{2} - c\frac{\theta_{x,A}}{2}s\frac{\theta_{y,A}}{2}s\frac{\theta_{z,A}}{2} \\ c\frac{\theta_{x,A}}{2}s\frac{\theta_{y,A}}{2}c\frac{\theta_{z,A}}{2} + s\frac{\theta_{x,A}}{2}c\frac{\theta_{y,A}}{2}s\frac{\theta_{z,A}}{2} \\ c\frac{\theta_{x,A}}{2}c\frac{\theta_{y,A}}{2}s\frac{\theta_{z,A}}{2} - s\frac{\theta_{x,A}}{2}s\frac{\theta_{y,A}}{2}c\frac{\theta_{z,A}}{2} \end{bmatrix}, \tag{1.16}$$

and comparing the results to Q_D's calculation, we obtain

$$q_{2,0} = -q_{1,1}, q_{2,1} = q_{1,0}, q_{2,2} = q_{1,3}, q_{2,3} = -q_{1,2}. \tag{1.17}$$

1.3.4 Assistive Follower Robot (AFR)'s Position

In Fig. 1.7, AFR's EE is supposed to be placed in an adjustable artificial potential field that contains three sub-fields: $U_1(P_1, X_A)$, $U_2(P_2, X_A)$ and $U_3(P_3, X_A)$. They provide repulsive or attractive forces acting on AFR: $U_1(P_1, X_A)$ provides a repulsive force F_1 which avoids collision with the object; $U_2(P_2, X_A)$ generates an attractive force F_2 which guides AFR to a general ready-to-hold pose; and $U_3(P_3, X_A)$ produces an attractive force F_3 which modifies AFR's pose according to DFR's movement. Then, the total field is

$$U(X_A) = w_{u1}U_1(P_1, X_A) + w_{u2}U_2(P_2, X_A) + w_{u3}U_3(P_3, X_A), \tag{1.18}$$

where X_A represents the current position of AFR's EE, w_{u1}, w_{u2} and w_{u3} are the adjustable weights, and P_1, P_2 and P_3 are the measured object's position X_o (deviates from the real centre), the general ready-to-hold position based on the object's movement and the specific ready-to-hold position based on DFR's movement, respectively. P_2 is different from P_3 because the measured object's position is usually not on the holding line, and P_3 is computed via the operator's expected holding position while P_2 is not.

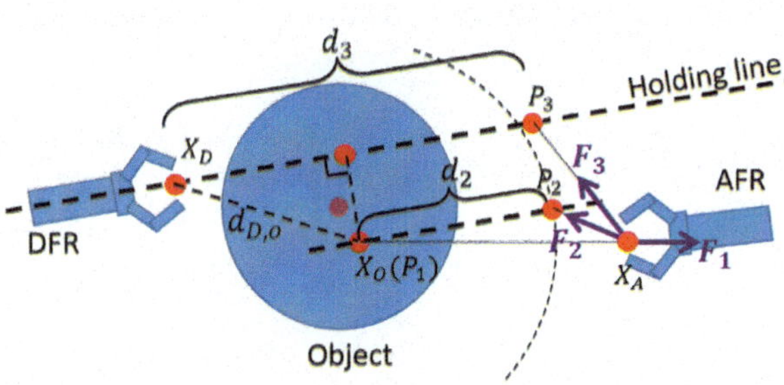

Fig. 1.7 The adjustable artificial potential field to regulate the movement of AFR

P_1, P_2 and P_3 are determined by

$$
\begin{aligned}
&P_1 = X_o, \quad P_2 = X_o + {}^F R_{D,t}[0, 0, d_2]^T, \\
&P_3 = X_1 + {}^F R_{D,t}[0, 0, d_3]^T,
\end{aligned}
\tag{1.19}
$$

where d_2 is an approximate distance according to objects' width that can be detected via machine vision and d_3 is a variable:

$$
d_3 = \begin{cases} \overline{\rho_{D,o}}, & \text{if } \ d_{D,o} > \overline{\rho_{D,o}} \\ \varpi d_{D,o}, & \text{otherwise} \end{cases}, \tag{1.20}
$$

where $d_{D,o}$ is the distance between DFR and the measured object's position, $\overline{\rho_{D,o}}$ is the upper bound to determine whether $d_{D,o}$ takes effect or not, i.e., P_3 takes effect or not based on the adjustable weights in (1.23), and $\varpi \leq 2$ ensures AFR to come closer enough to DFR to contact the object, generate contact force and trigger holding.

Then, three distances ($\rho_m = ||X_2 - P_m||_2$, $m = 1, 2, 3$) from P_m to X_2 are used to determine three sub-fields:

$$
\begin{aligned}
&U_1 = \begin{cases} \frac{1}{2}\eta_1(\frac{1}{\rho_1} - \frac{1}{\rho_{1,0}})^2, & \text{if } \ \rho_1 \leq \rho_{1,0} \\ 0, & \text{otherwise} \end{cases}, \\
&U_2 = \frac{1}{2}\eta_2\rho_2^2, \quad U_3 = \frac{1}{2}\eta_3\rho_3^2,
\end{aligned}
\tag{1.21}
$$

where $\eta_m, m = 1, 2, 3$ are constant gains and $\rho_{1,0}$ is set by the operator and determines the distance boundary in which U_1 takes effect. The resultant force acting on AFR is

$$
\begin{aligned}
F &= -\text{grad}[U] = w_1 F_1 + w_2 F_2 + w_3 F_3 \\
&= w_1\eta_1(\frac{1}{\rho_1} - \frac{1}{\rho_{1,0}})\frac{1}{\rho_1^2}\frac{\partial \rho_1}{\partial X} - w_2\eta_2\rho_2\frac{\partial \rho_2}{\partial X} - w_3\eta_3\rho_3\frac{\partial \rho_3}{\partial X}.
\end{aligned}
\tag{1.22}
$$

where U is the total artificial potential field represented in (1.18). For the adjustable weights, they are affected by $d_{D,o}$:

$$
\begin{aligned}
&w_{u1} = max\left\{0, min\left\{1, \frac{Pro_{d_{D,o}} - \underline{\rho_{D,o}}}{\overline{\rho_{D,o}} - \underline{\rho_{D,o}}}\right\}\right\}, \\
&w_{u2} = w_{u1}, \quad w_{u3} = 1 - w_{u1},
\end{aligned}
\tag{1.23}
$$

where $\underline{\rho_{D,o}}$ is the lower bound for computing weights and $Pro_{d_{D,o}}$ is the projection of $d_{D,o}$ in the holding line. These time-varying weights contribute to a smooth adjustment of the field $U(X_2)$: when DFR is far from the object, only U_1 and U_2 play the role in regulating AFR's motion; when DFR gets close to the object, U_3 gradually takes over and the effect of U_1 and U_2 fades away.

Finally, we have the updated AFR position

$$X_{A,t} = X_{A,t-\Delta t} + \Delta X_A \tag{1.24}$$

with $\Delta X_A = sgn(F) * min\{|k_p F|, \Delta X_{A,max}\}$, where Δt is the sampling time and ΔX_A is the increment displacement for AFR due to the field force F, with k_p standing for a proportional coefficient that transfers a force into a displacement, and $\Delta X_{A,max} > 0$ limits the AFR's moving speed.

1.4 Motion Regulation During Object Holding

This section corresponds to the second contribution of the section, which mainly contains the motion regulation for AFR to assist in holding a deformable object with the desired force. Based on (1.19) and (1.20), P_3 will finally go "inside" the object and AFR will get in contact with the object. The contact force is a signal to invalidate the artificial field and to initialize the regulation when holding the object. The objective is to generate a reference trajectory for AFR in order to maintain a desired contact force, which is an extension of our previous work [26]. As the previous sections have aligned the two followers' z axes in one line ($\overline{X_D P_3}$), most of the variables and calculations discussed in this section are one-dimensional along this line. Thus, we leave out the left-side superscripts/subscripts for brevity.

We depict a general case in Fig. 1.8, where two followers hold an object and AFR's current contact force f_z is not equal to the desired contact force f_d which corresponds to an unknown position z_d. We model the object as a spring. And thus, we have

$$(-f_z) - (-f_d) = k_o(z_A - z_d) = -e_f, \tag{1.25}$$

where $e_f = f_z - f_d$ is the force tracking error and k_o is the unknown elastic coefficient. For AFR, we consider the contact with the object to satisfy an impedance model. As our purpose is to derive a trajectory z_r for AFR to track such that $e_f \to 0$, it satisfies

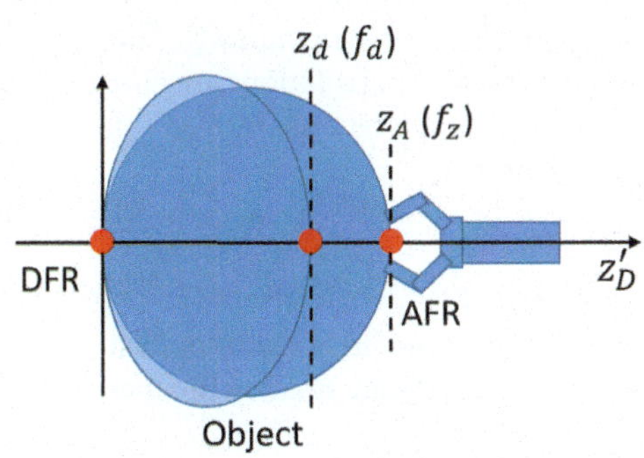

Fig. 1.8 A sketch to show the relationship of the deformable object, DFR, AFR, the z-axis positions in DFR's tool frame and the corresponding forces of AFR

$$M\ddot{e}_z + D\dot{e}_z + Ke_z = e_f, \tag{1.26}$$

where $e_z = z_A - z_r$, M, D and K represent positive impedance model parameters.

As z_d is a desired continuous trajectory, based on the autoregressive model, we can model it by

$$\dot{z}_d(t) = w_0 + w_1 z_A(t) + w_2 z_A(t - \Delta t) + w_m z_A(t - (m-1)\Delta t), \tag{1.27}$$

which can be rewritten as

$$\dot{z}_d = W_z^T Z, \tag{1.28}$$

where $W_z = [w_0, w_1, w_2, \cdots, w_m]^T$ is the weight and $Z = [1, z_A(t), z_A(t - \Delta t), \cdots, z_A(t - (m-1)\Delta t)]$. We approximate $\dot{z}_d$ via

$$\dot{\hat{z}}_d = \hat{W}_z Z - L\tilde{z}_d(t - \Delta t), \tag{1.29}$$

where $\dot{\hat{z}}_d$ and $\hat{W}_z$ are the approximations of the desired velocity and W_z, and L is a positive constant. The updating law for $\hat{W}_z$ is designed as

$$\dot{\hat{W}}_z = -(\tilde{z}_d(t - \Delta t) + \beta e_f)Z(t), \tag{1.30}$$

where β is a positive constant and

$$\begin{aligned} \tilde{z}_d(t - \Delta t) &= \hat{z}_d(t - \Delta t) - z_d(t - \Delta t) \\ &= \hat{z}_d(t - \Delta t) - z_A(t). \end{aligned} \tag{1.31}$$

As AFR's position z_A always tracks z_d, we have $z_d(t - \Delta t) = z_A(t)$. Then, the reference trajectory z_r for AFR is

$$\dot{z}_r = \dot{\hat{z}}_d - \alpha e_f + L\tilde{z}_d(t - \Delta t), \tag{1.32}$$

where α is a positive constant. To generate the reference trajectory z_r, the updating law (1.32) is based on three parts: $\dot{\hat{z}}_d$ is the estimation of the desired contact position; $-\alpha e_f$ is a correction based on contact force feedback; and $L\tilde{z}_d(t - \Delta t)$ is a compensation for the coupling effect.

As the above calculations are along the z'_D axis of DFR's tool frame, the reference trajectory should be transformed into the follower's frame via

$$X_{A,t} = X_{D,t} + {}^F R_{D,t}[0, 0, z_r]^T. \tag{1.33}$$

With (1.33), we can simplify the problem to find AFR's 3D position into a problem to determine its 1D holding distance. For convergence analysis, we construct a Lyapunov candidate to involve all errors:

$$V = V_1 + V_2 + V_3 + V_4, \tag{1.34}$$

where

$$\begin{aligned} V_1 &= \frac{1}{2\beta}\tilde{W}_z^T\tilde{W}_z, \quad V_2 = \frac{1}{2\beta}\tilde{z}_d^2 \\ V_3 &= \frac{1}{2}(M\dot{e}_z^2 + Ke_z^2), \quad V_4 = \frac{1}{2k_o}e_f^2. \end{aligned} \tag{1.35}$$

with $\tilde{W}_z = W_z - \hat{W}_z$. Based on system dynamics (1.25) and (1.26), the derivation of (1.34) with respect to time is

$$\dot{V} = \dot{V}_1 + \dot{V}_2 + \dot{V}_3 + \dot{V}_4, \tag{1.36}$$

with

$$\begin{aligned} \dot{V}_1 &= \frac{1}{\beta}\tilde{W}_z^T\dot{\tilde{W}}_z, \quad \dot{V}_2 = \frac{1}{\beta}\tilde{z}_d\dot{\tilde{z}}_d, \\ \dot{V}_3 &= \dot{e}_z(M\ddot{e}_z + Ke_z) = \dot{e}_z(e_f - D\dot{e}_z), \\ \dot{V}_4 &= \frac{1}{k_o}e_f\dot{e}_f = e_f(\dot{z}_2 - \dot{z}_d). \end{aligned} \tag{1.37}$$

As $\dot{W}_z = 0$, we have

$$\dot{V}_1 = \frac{1}{\beta}\tilde{W}_z^T\dot{\hat{W}}_z = -\tilde{W}_z^T(\frac{1}{\beta}\tilde{z}_d(t-\Delta t) + e_f)Z(t). \tag{1.38}$$

Subtracting (1.28) from (1.29), we obtain

$$\dot{\tilde{z}}_d(t) = \dot{\tilde{W}}_z Z(t) - L\tilde{z}_d(t-\Delta t). \tag{1.39}$$

Then,

$$\dot{V}_2 = \frac{1}{\beta}\tilde{z}_d(\dot{\tilde{W}}_z Z(t) - L\tilde{z}_d(t-\Delta t)). \tag{1.40}$$

Assuming Δt is small, we acquire $\dot{\tilde{z}}_d(t) \to \tilde{z}_d(t-\Delta t)$ and thus

$$\dot{V}_1 + \dot{V}_2 = -e_f\tilde{W}^T Z - \frac{1}{\beta}L\tilde{z}_d^2. \tag{1.41}$$

With $\dot{\tilde{z}}_d = \dot{\hat{z}}_d - \dot{z}_d$, employing the reference trajectory updating law (1.32), we have

$$\dot{z}_d = \dot{\hat{z}}_d - \dot{\tilde{z}}_d = \dot{z}_r + \alpha e_f - L\tilde{z}_d(t-\Delta t) - \dot{\tilde{z}}_d. \tag{1.42}$$

Substituting (1.42) into $\dot{V}_4$ and resorting to (1.39), we obtain

$$\begin{aligned}\dot{V}_4 &= e_f(\dot{z}_A - (\dot{z}_r + \alpha e_f - L\tilde{z}_d(t-\Delta t) - \dot{\hat{z}}_d)) \\ &= e_f\dot{e}_z - \alpha e_f^2 + e_f\tilde{W}_z^T Z.\end{aligned} \tag{1.43}$$

Finally, substituting (1.41) and (1.43) into (1.36), we have

$$\dot{V} = -\frac{1}{\beta}L\tilde{z}_d^2 - D\dot{e}_z^2 - \alpha e_f^2 \leq 0, \tag{1.44}$$

which means $\tilde{z}_d$, $\dot{e}_z$ and e_f converge when $t \to \infty$. As a result, it is worth pointing out that we have $f_z \to f_d$ with e_f converging, namely the desired contact force is maintained.

1.5 Simulation and Experiment Study

We perform simulations to verify the contact force regulation and experiments to demonstrate the whole proposed solution.

1.5.1 Simulations

As depicted in Fig. 1.9, the simulation is performed on a platform with two 3-DOF manipulators in MATLAB, where the two robots hold a deformable object with time-varying parameters. The bases of the two robots are DFR:(0,0) and AFR:(0.5,0). Their link lengths from link 1 to link 3 are 0.5, 0.3 and 0.1. Their initial joint positions from base to EE are DFR($2\pi/3, -\pi/3, -\pi/3$) and AFR($\pi/3, \pi/3, \pi/3$). This results in the two robots just touching the surface of the object without generating any force. Note that we use SI units throughout Sect. 1.5 and thus leave out units for brevity. As shown in Fig. 1.10, three cases are simulated and each column represents one case. Two teleoperation trajectories for DFR are employed: $x_D = min(0.35, 0.015t)$ in the first and the second columns and $x_D = 0.2sin(0.3t)$ in the third column. The y components are set to the same constant as its initial value. The elastic coefficient of the object is set as a constant $k_o = 50$ in the first column and a variable $k_o = 50 + 10sin(0.5t)$ in the second and the third columns, which are unknown to the system. The desired contact force is $f_d = 10$ in all cases. The reference trajectories for AFR are generated via the method developed in Sect. 1.4. The initial weights are set to be $W_z = [0, 0, 0, 0]^T$, $L = 2$, $\alpha = 0.11$, $\beta = 0.0001$ and $x_r(0) = \dot{x}_r(0) = \hat{x}_d(0) = 0$. The proposed method is compared with two other controllers: the force controller $\Delta x_D = -0.00005e_f$ and the passive-impedance based controller $6\ddot{e}_{x_D} + 2\dot{e}_{x_D} = e_f$.

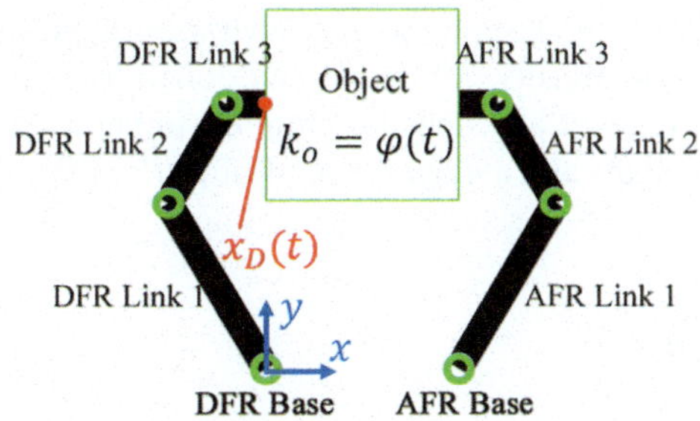

Fig. 1.9 The simulation is performed on a platform with two 3-DOF manipulators. Two followers are commanded to manipulate a deformable object, where DFR tracks predefined teleoperating trajectories, AFR regulates its motion and the object's physical parameter is time-varying

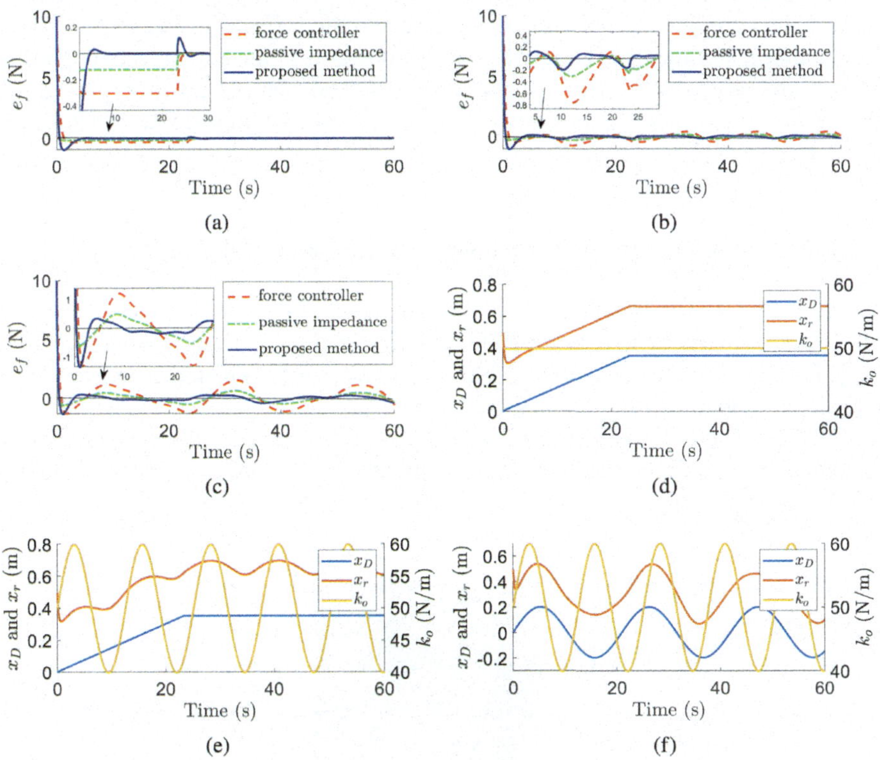

Fig. 1.10 The simulation results. The first row: force tracking errors by three methods. The second row: updating system states x_d, x_r and k_o. Each column shares one set of system parameters illustrated in the second row

As seen in the first row of Fig. 1.10, the proposed method performs the best among the three controllers with less force tracking error during most time periods. In the second row, we see that reference trajectories for AFR are generated, leading to the desired contact force, despite the unknown movement of DFR and different unknown parameters of objects. Thus, we can conclude that the proposed force controller performs robustly and depends little on unknown factors including the random DFR's movements and a large range of object's stiffness (40–60 N/m).

1.5.2 Experiments

The experimental configuration is depicted in Fig. 1.11, where two arms of the Baxter robot are used as two follower robots and the Touch X is the leader robot. The Touch X has a built-in button on the joystick, which is used to trigger re-"fixing" technique for pose transformation. To position the object, we resort to an RGB-D camera Kinect V1 and the coordination between the camera and the followers is calibrated before conducting experiments. The whole system runs at a frequency of 100 Hz and the images are processed at 15 frames per second. The object's position is obtained via calculating the average position ($\frac{1}{4}\sum_{n=1}^{4} O_{En}$) of four edge points illustrated in Fig. 1.12.

1.5.2.1 Demonstration on Motion Regulation Before Holding

There are two key points that are worth demonstrating before holding. One is the effectiveness of the relative pose transformation technique. The other is the regulation of AFR, including the adjustable APFM to regulate AFR's position and the practical technique to obtain AFR's orientation.

The relative pose transformation technique largely improves the ergonomics and the precision of the teleoperation, which is intuitively demonstrated by Fig. 1.13 and Fig. 1.14.

Fig. 1.11 The experimental configuration. The right arm and the left arm of the Baxter robot are set as DFR and AFR respectively. Touch X is employed as the leader. A Kinect RGB-D camera is applied to locate the object's position. Images are processed on a Windows computer and other calculations are on a Linux computer. Two target positions are marked on the table in the followers' task space

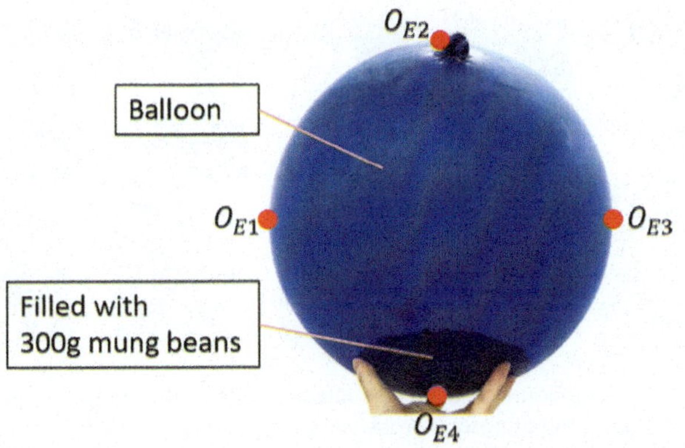

Fig. 1.12 A balloon is treated as the object to be held and it is filled with 0.3kg mung beans constituting 3N gravity. The average of the four edge points' positions determines the object's position

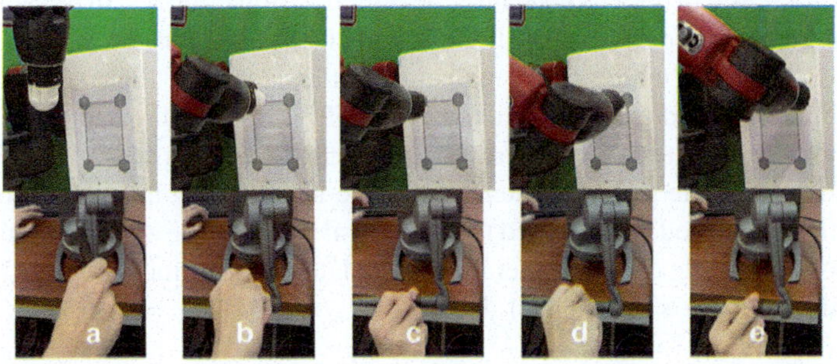

Fig. 1.13 Task 1: An uncomfortable and imprecise teleoperation example, mimicking the method of matching the whole workspace. **a** initial pose; **b** working pose; **c** uncomfortably tighten/unscrew a nut; **d** hard to precisely move to other nut's position; **e** uncomfortably tighten/unscrew another nut

When the operator teleoperates DFR to tighten/unscrew two nuts, in Fig. 1.13, the operator can hardly control DFR to precisely move from one nut to the other, and the operator has to do an uncomfortable (unergonomic) gesture to tighten/unscrew a nut; while in Fig. 1.14, the operator uses a smaller scaling factor and the re-"fixing" technique, he/she can perform precise teleoperation and use any comfortable (ergonomic) gesture as he/she desired to tighten/unscrew a nut.

To be specifically, comparing Figs. 1.13c and 1.14c, we can see that when tightening/unscrewing the same nut, the proposed method provides a more ergonomic operating pose. As the manipulator's (Baxter's) workspace is much larger than a desktop device's (Touch X's), when the follower needs to move to another position, the operator finds it hard to precisely control the leader (Fig. 1.13d). By setting a smaller scaling coefficient in the relative pose transformation method, we can achieve accurate teleoperation (Fig. 1.14d). Obviously, if the scaling factor is set to a larger value, we can also achieve a flexible teleoperation.

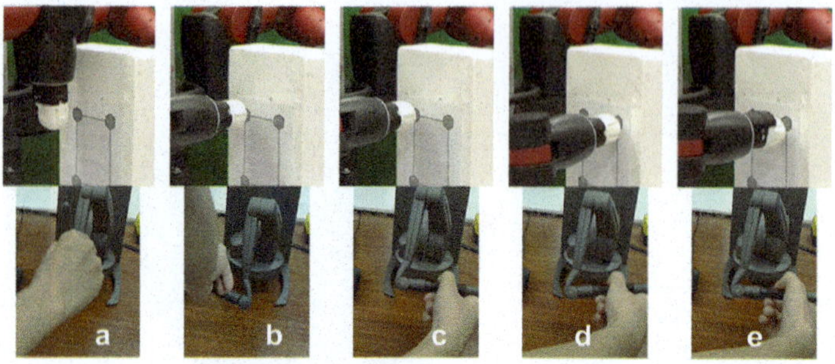

Fig. 1.14 Task 1: An ergonomic and precise teleoperation based on the proposed relative pose transformation method. **a** Initial pose; **b** working pose; **c** working pose but the operator can re-set an ergonomic initial operating pose to tighten/unscrew a nut; **d** easier to accurately move to another nut's position; **e** comfortably tighten/unscrew another nut

We do not really drive a nut because the pure position control can not achieve that, whereas the general compliant control can not clearly demonstrate the advantage of the proposed method which performs better in the aspect of tele-manipulation precision. Thus, we conduct another experiment (Fig. 1.15a): to teleoperate the DFR to perpendicularly hit six dartboard centers one by one on a wall, using marker pens (blue pen for the general match-the-whole-space method, red pen for the proposed re-fixing method). Compared to the general method, the adaptive scaling factor s facilitates precise/flexible teleoperation. It updates in real time based on the moving speed of the operator's hand via (1.45)

$$s = \begin{cases} s_b, & if \quad v < v_{min} \\ s_b + \dfrac{min(v, v_{max}) - v_{min}}{v_{max} - v_{min}} * s_p, & otherwise \end{cases} \tag{1.45}$$

where $s_b = 1$ is a basic scaling factor, v is the moving speed of the leader, $v_{max} = 0.3$ and $v_{min} = 0.05$ are predefined upper and lower bounds of moving speeds and $s_p = 7$ is a linear transform to in-/decrease s. For instance, the faster the hand moves ($v \leq v_{max}$), the larger the s becomes.

The experimental results are shown in Fig. 1.15. To complete this task, when the DFR gets close to a dartboard center, it slows down and tries to precisely hit the center (e.g., Fig. 1.15b). This can be achieved only if the operator slows down his/her hand's movement and gently teleoperates the DFR. When the DFR moves from one dartboard center to another (e.g., Fig. 1.15c), the operator can move faster as this movement requires less precision. The faster hand movement contributes to the DFR moving faster and longer than the operator's hand's displacement. Therefore, the adjustable s facilitates the precise/flexible teleoperation.

We ask 13 persons (3 females, 10 males; aged from 22 to 30) to conduct this experiment via the general match-the-whole-space method and the proposed re-fixing and adaptive-scaling method. We record the hit points with a blue pen and

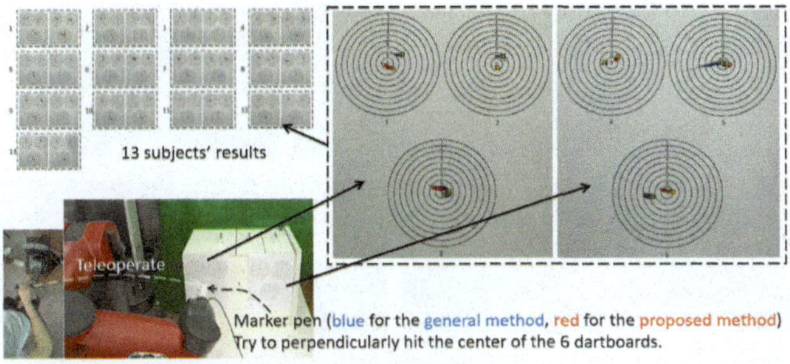

(a) A task to teleoperate DFR to use marker pens to perpendicularly hit 6 dartboard centers.

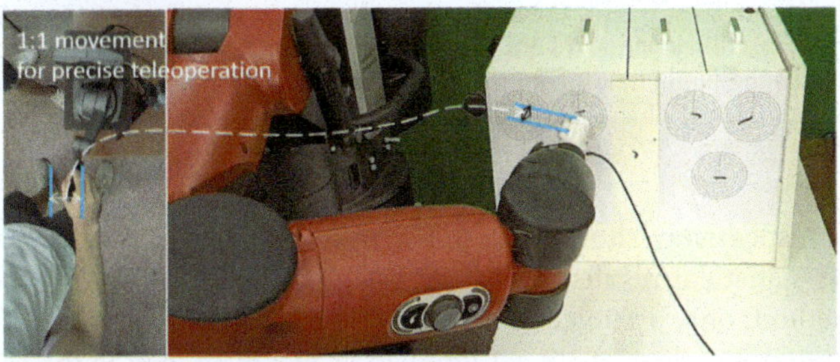

(b) When DFR gets close to a dartboard center, the operator slows down and thus s becomes smaller in order to conduct precise teleoperation.

(c) When precision is not important, the operator can move faster and thus s increases to boost DFR's moving speed.

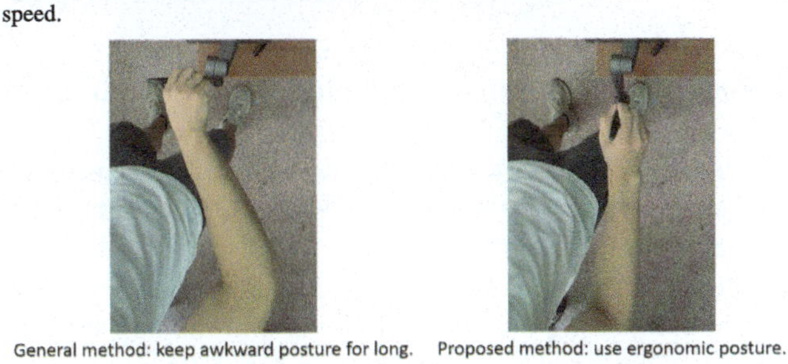

(d) The proposed re-fixing method enables the operator to use any posture to conduct teleoperation.

Fig. 1.15 Task 2: The experiment of teleoperating DFR to hit dartboard centers with marker pens

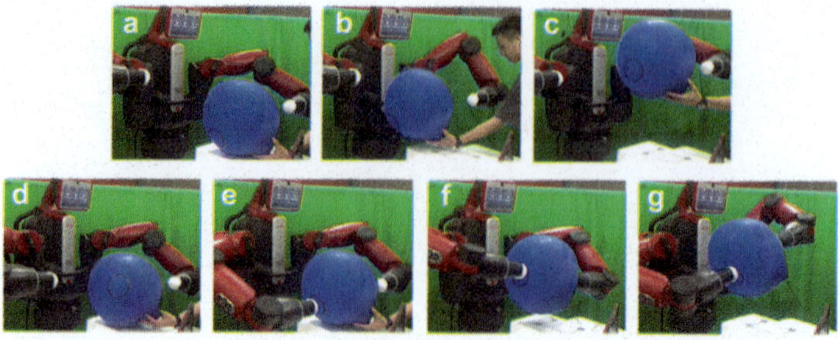

Fig. 1.16 Task 3: A demo on regulating AFR's (Baxter's left arm's) pose. **a** AFR tracks the object that moves left-/rightwards; **b** AFR tracks the object that moves for-/backwards; **c** AFR tracks the object that moves up-/downwards; **d** DFR starts to track the moving target holding point (the black circle); **e** DFR hits the target holding point and AFR assists in holding at the right time; **f**, **g** AFR keeps assisting holding the object in a facing pose regardless of DFR's rotations

a red pen to respectively show the results by the general method and the proposed method. The statistical results show that: (i) 12 subjects get higher scores via the proposed method, only 1 subject gets the same scores via the two methods; (ii) all subjects mention that they easily get fatigued or feel even pain in the forearm when using the general method (as it required him/her to keep a pronation posture for a long time), but feel relaxed when using the proposed re-fixing method.

Consequently, we can conclude that our method facilitates flexible/precise and ergonomic teleoperation.

The demonstration of regulating AFR's pose is illustrated in Fig. 1.16, where AFR (Baxter's left arm) tracks the object and also avoids collision when the object moves left-/rightwards (a), for-/backwards (b) and up-/downwards (c); and AFR regulates its motion to a ready-to-hold pose (d and e) when DFR moves close to the object and the target holding point. The efficient quaternion-based technique to obtain AFR's holding orientation is also verified as shown in d, e, f and g, where we can see that AFR always keeps pointing to DFR.

1.5.2.2 Demonstration on Motion Regulation During Holding

The first is shown in Fig. 1.17, we use the SLDF system to transport general rigid objects with different sizes: a hardboard box with different side lengths - holding different sides every time to pretend to hold different objects. With the help of the proposed motion regulation solutions for AFR, the AFR successfully and autonomously assists the DFR in completing the task of holding and moving the object from one position into a box. Combined with the demo in Fig. 1.19, intuitively, as the force controller is devised based on an impedance model, we can tell that the proposed method is capable of compliantly transporting different volumetric objects, no matter whether rigid or deformable, small or large in size.

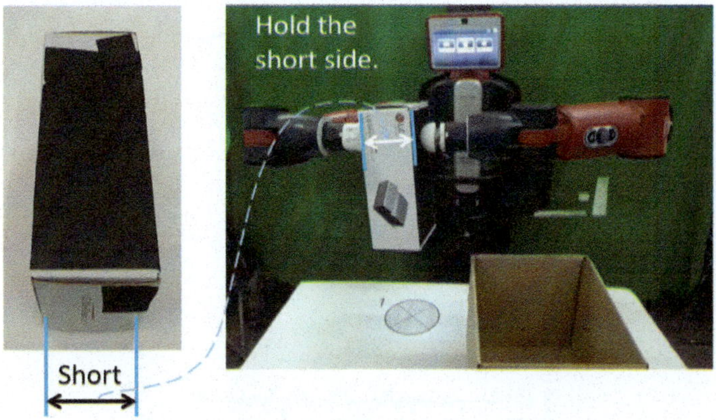

(a) Transport a rigid small-size object.

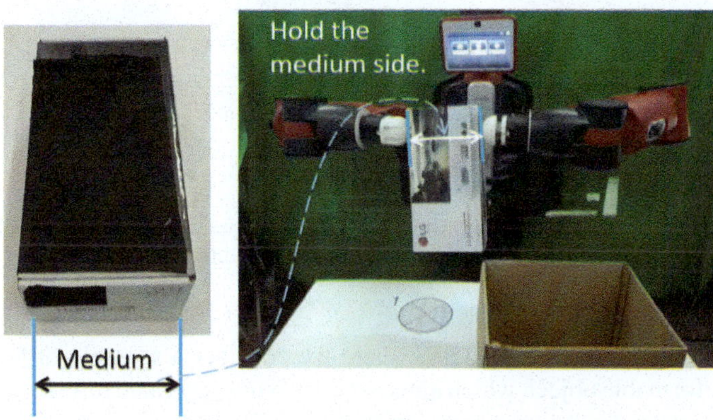

(b) Transport a rigid medium-size object.

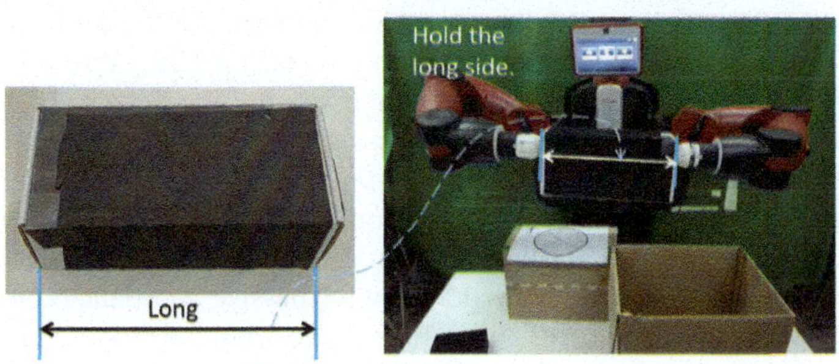

(c) Transport a rigid large-size object.

Fig. 1.17 Task 4: The experiment of holding and moving rigid objects with different sizes. A hardboard box with different lengths of its three sides is used to mimic three rigid objects. The operator teleoperates DFR. AFR autonomously assists in holding the object, moving it and dropping it into a box

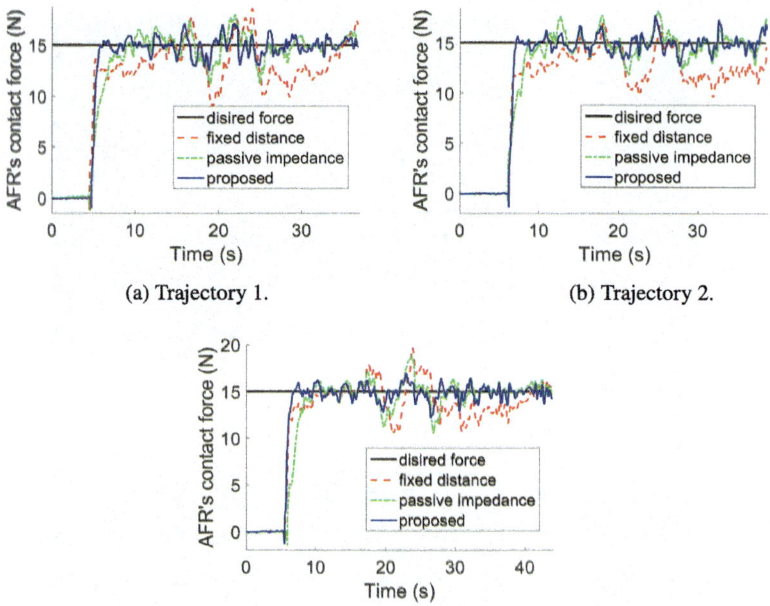

(a) Trajectory 1.

(b) Trajectory 2.

(c) Trajectory 3.

Fig. 1.18 Comparison of AFR's contact force tracking performance when applying three teleoperated trajectories and implementing three holding methods

The second is shown in Fig. 1.19, where we consider a scenario of holding and moving a deformable object, which aims to verify if the generated reference trajectory helps to maintain the contact force within a small range despite DFR's random movement and the object's time-varying internal force. As seen in the caption of Fig. 1.19, this task contains linear and angular acceleration of the held object, which can cover most real-world hold-and-move tasks. Note that during this task, AFR will suffer from irregular time-varying contact force.

The operator executes this task three times and three teleoperation trajectories are recorded, which are used for comparing three holding methods: fixed-distance holding; holding with a passive impedance controller to adjust AFR's position; and holding with the proposed controller to adjust AFR's position. For the fixed-distance holding, the holding distance is set as 0.38. For the passive impedance controller, the passive impedance model is set as $10\ddot{e}_{z_D} + 2\dot{e}_{z_D} = e_f$. For the proposed method, the initial parameters are set to be $W_z = [0, 0, 0]^T$, $L = 50$, and $\alpha = 0.25$, $\beta = 0.0001$ and $z_r(0) = \dot{z}_r(0) = \hat{z}_d(0) = 0$. $f_d = 15$. Results of the comparisons are shown in Fig. 1.18.

Obviously, the proposed method performs the best in force tracking, being the most stable and the closest to the desired force regardless of different teleoperation trajectories and the object's varying internal force conditions. The force tracking errors are the smallest when using the proposed method, with the root mean square

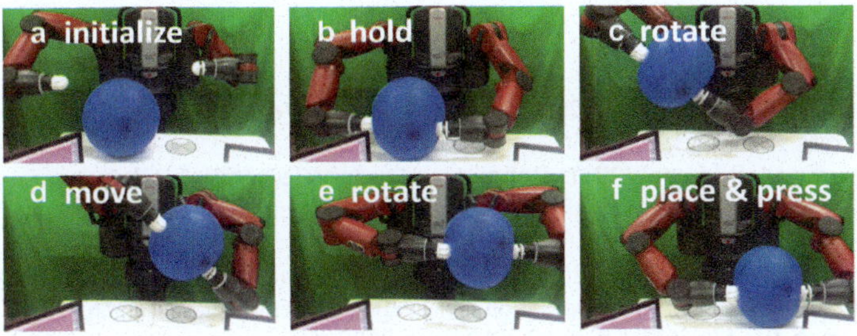

Fig. 1.19 Task 5: A deformable object hold-and-move scenario in an SLDF teleoperation. **a** initial positions of the object, DFR and AFR; **b** followers collaboratively hold the object on the left target position; **c** followers rotate the held object by $\pi/4$ to let the object's gravity act on AFR; **d** followers keep the holding pose and move left-/rightwards with ac-/deceleration; **e** followers hold the object over the right target position with a normal pose; **f** followers hold the object and keep the object press on the right target position for three seconds

errors (RMSE) (blue, green, red; after the 8th second): 0.87, 1.22, 2.48 in Fig. 1.18a; 0.82, 1.21, 2.61 in Fig. 1.18b; 0.77, 1.31, 1.93 in Fig. 1.18c. Besides, the proposed method is the fastest way to reach the desired force with neglectable overshoot, which in other aspects demonstrates that the adaption law in (1.29) converges fast. These results validate that with the design of the force controller targeted on this kind of hold-and-move task, our method can best adjust AFR's motion to maintain the desired force, outperforming the general passive-impedance model-based force controller and the position-based controller.

It is also worth mentioning that, the quaternion-based technique to obtain AFR's orientation is also employed in this experiment, and from the frames shown in Fig. 1.19, we can see that with this technique, AFR can assist in holding and rotating the held object without causing any twist/torque to the object.

1.6 Conclusion

This chapter provides a set of solutions for an SLDF teleoperation system to hold and move a deformable object. It includes the kinematic solutions to deriving the relative DFR's pose and AFR's ready-to-hold pose, where we present a technique of relative pose transformation, the practical quaternion-based solution to obtain AFR's orientation that assists in holding and the adjustable APFM to regulate AFR's position. Besides, based on the autoregressive model and impedance model, the force controller is designed to ensure that AFR adjusts itself to the right position to maintain a desired force when AFR collaborates with DFR to hold and move a deformable object. Simulations and experiments are conducted, verifying the feasibility and effectiveness of the proposed methods.

References

1. Shahbazi M, Atashzar SF et al (2018) A systematic review of multilateral teleoperation systems. IEEE Trans Hapt 11(3):338–356
2. Feizi N, Tavakoli M et al (2021) Robotics and ai for teleoperation, tele-assessment, and tele-training for surgery in the era of covid-19: existing challenges, and future vision. Front Robot AI 8:610677
3. Panzirsch M, Singh H et al (2020) Safe interactions and kinesthetic feedback in high performance earth-to-moon teleoperation. In: Paper presented at the 2020 IEEE Aerospace Conference, Big Sky, MT, USA, 07–14 March 2020
4. Sun D, Liao Q et al (2020) Single master bimanual teleoperation system with efficient regulation. IEEE Trans Robot 36(4):1022–1037
5. Huang D, Yang C et al (2020) Disturbance observer enhanced variable gain controller for robot teleoperation with motion capture using wearable armbands. Autonom Robot 44(7):1217–1231
6. Rakita D, Mutlu B et al (2019) Shared control–based bimanual robot manipulation. Sci Robot 4(30):eaaw0955
7. Yang Y, Constantinescu D et al (2019) Passive multiuser teleoperation of a multirobot system with connectivity-preserving containment. IEEE Trans Robot 38(1):209–228
8. Huang P, Dai P et al (2018) Asymmetric wave variable compensation method in dual-master-dual-slave multilateral teleoperation system. Mechatronics 49:1–10
9. Li J, You B et al (2022) Dual-master/single-slave haptic teleoperation system for semiautonomous bilateral control of hexapod robot subject to deformable rough terrain. IEEE Trans Syst Man Cybern Syst 52(4):2435–2449
10. Sun D, Liao Q et al (2018) Multilateral teleoperation with new cooperative structure based on reconfigurable robots and type-2 fuzzy logic. IEEE Trans Cybern 49(8):2845–2859
11. Sun Y, Pan B et al (2021) Vision-based framework of single master dual slave semi-autonomous surgical robot system. IRBM 42(1):55–64
12. Watanabe K, Kanno T et al (2017) Single-master dual-slave surgical robot with automated relay of suture needle. IEEE Trans Indus Electron 65(8):6343–6351
13. Sanchez D, Wan W et al (2021) Four-arm collaboration: two dual-arm robots work together to manipulate tethered tools. IEEE/ASME Trans Mechatron 27(5):3286–3296
14. Alonso-Mora J, Knepper R et al (2015) Local motion planning for collaborative multi-robot manipulation of deformable objects. In: Paper presented at the 2015 IEEE International Conference on Robotics and Automation (ICRA), Seattle, WA, USA, 26–30 May 2015
15. Sanchez J, Corrales J et al (2018) Robotic manipulation and sensing of deformable objects in domestic and industrial applications: a survey. Int J Robot Res 37(7):688–716
16. Liu H, Liu HHT et al (2020) Navigation information augmented artificial potential field algorithm for collision avoidance in UAV formation flight. Aerospace Syst 3:229–241
17. Wang W, Zhu M et al (2018) An improved artificial potential field method of trajectory planning and obstacle avoidance for redundant manipulators. Int J Adv Robot Syst 15(5):1729881418799562
18. Khatib O (1986) Real-time obstacle avoidance for manipulators and mobile robots. Int J Robot Res 5(1):90–98
19. Luo L, Wen H et al (2021) An obstacles avoidance method for serial manipulator based on reinforcement learning and Artificial Potential Field. Int J Intell Robot Appl 5:186–202
20. Wan W, Harada K (2016) Developing and comparing single-arm and dual-arm regrasp. IEEE Robot Autom Lett 1(1):243–250
21. Caccavale F, Chiacchio P et al (2000) Task-space regulation of cooperative manipulators. Automatica 36(6):879–887
22. Yin H, Varava A et al (2021) Modeling, learning, perception, and control methods for deformable object manipulation. Sci Robot 6(54):eabd8803
23. Zimmermann S, Poranne R et al (2021) Dynamic manipulation of deformable objects with implicit integration. IEEE Robot Autom Lett 6(2):4209–4216

24. Howard AM, Bekey GA (2000) Intelligent learning for deformable object manipulation. Autonomous Robots 9:51–58
25. Lu Z, Huang P et al (2019) Relative impedance-based internal force control for bimanual robot teleoperation with varying time delay. IEEE Trans Indus Electron 67(1):778–789
26. Li Y, Yang C (2019) A hybrid human motion prediction approach for human-robot collaboration. In: Advances in Intelligent Systems and Computing. AISC, vol 1043. Springer
27. Chiacchio P, Chiaverini S et al (1991) Closed-loop inverse kinematics schemes for constrained redundant manipulators with task space augmentation and task priority strategy. Int J Robot Res 10(4):410–425
28. Yazdani A, Novin RS et al (2021) Is the leader robot an adequate sensor for posture estimation and ergonomic assessment of a human teleoperator? In: Paper presented at the 2021 IEEE 17th international conference on automation science and engineering (CASE), Lyon, France, 23–27 August 2021
29. Mukhopadhyay P, O'Sullivan L et al (2007) Estimating upper limb discomfort level due to intermittent isometric pronation torque with various combinations of elbow angles, forearm rotation angles, force and frequency with upper arm at 90 abduction. Int J Indus Ergonom 37(4):313–325
30. O'Sullivan LW, Gallwey TJ et al (2005) Forearm torque strengths and discomfort profiles in pronation and supination. Ergonomics 48(6):703–721
31. Diebel J (2006) Representing attitude: Euler angles, unit quaternions, and rotation vectors. Matrix 58(15–16):1–35

Chapter 2
Single-Leader-Dual-Follower Cooperative Manipulation of Deformable Objects

2.1 Introduction

Different from the context in Chap. 1, in this chapter, we will research deformable object manipulation using single-leader-dual-follower (SLDF) robot teleoperation. Many SLDF works are about multiple mobile robots [1, 2], manipulators are the main focus of this chapter, where their cooperation is achieved via physical contact. Sun et al. [3] investigated the regulation of dual tele-robots' position, orientation and force in SLDF teleoperation. Sun et al. [4] developed a vision-based framework, enabling one tele-robot to semi-autonomously assist a tele-robot in performing minimally invasive surgery. Watanabe et al. [5] developed an approach for some ubiquitous surgical tasks, e.g., suturing, by using SLDF teleoperation.

The aforementioned SLDF teleoperation literature has an assumption that the follower robots can share information, e.g., their poses, with each other. However, there are many cases where AFR needs to cooperate with DFR, but they do not have direct communication. As shown in Fig. 2.1, while DFR is directly tele-operated by a human, AFR needs to modulate its own movement for successful manipulation of the object. Since these two follower robots are in physical interaction through a common object, DFR's movement may be estimated by AFR using the information of interaction force. Moreover, as for manipulation of deformable objects such as organs, tissues, cables and household laundries [6–8], researchers presented practical methods to endow the manipulator with touching sense [9] and achieve high-precision force tracking when interacting with soft objects [10, 11]. These works shared the same idea that a robot can use the contact force to modulate its movement. They also revealed challenges in addressing the problems when the objects' information is entirely unknown.

To address all the concerns mentioned-above, the proposed method has its unique properties for followers' collaboration in manipulating a deformable object, listed below:

C. Yang et al., *Advanced Teleoperation and Robot Learning for Dexterous Manipulation*, Springer Tracts in Advanced Robotics 160,
https://doi.org/10.1007/978-3-031-78501-6_2

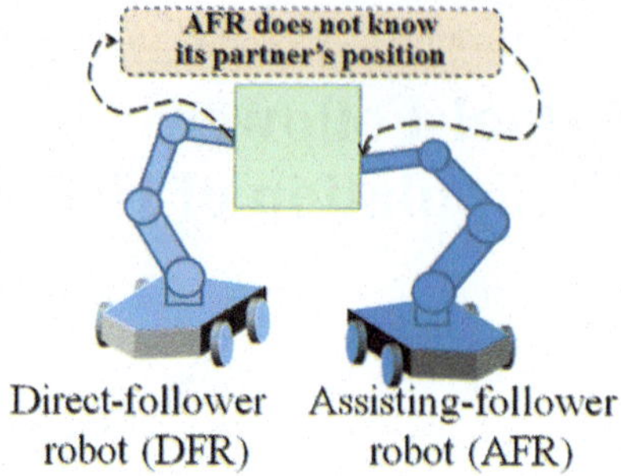

Fig. 2.1 An AFR needs to manipulate an object in collaboration with a DFR whose movement is controlled by a human operator and unknown to AFR

- Comparing to previous works [3–5], where both tele-robots exchange pose information with each other, two tele-robots do not know their partner's pose and their collaboration is achieved based on merely the contact force. Different from traditional passive impedance control [12–15], we enable AFR to estimate its partner's movement, leading to an explicit contact force control.
- Comparing to [16] and its extensions like [17–19], instead of dealing with an environment with a fixed rest position, the proposed method enables AFR to interact with an environment with a time-varying rest position (because the contact surface moves as DFR moves), without requiring the knowledge of the environmental movements and interaction model parameters.

2.2 Problem Formulation

2.2.1 System Description

The system setup to be discussed in this section is shown in Fig. 2.2, where the leader robot is manipulated by an operator, and its position $X_l = [x_l, y_l, z_l]^T$ is conveyed to DFR, and AFR assists DFR in holding and moving a deformable object.

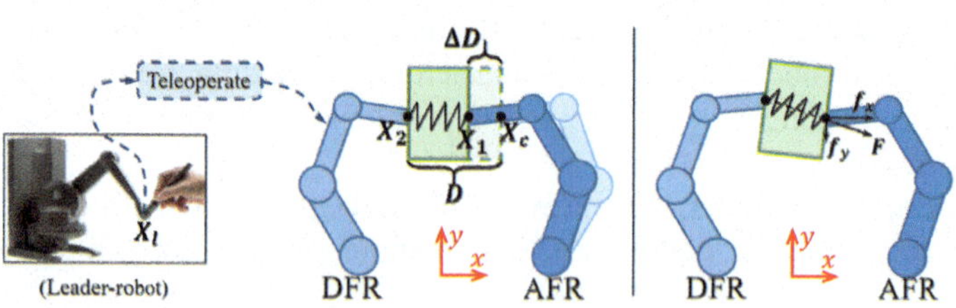

Fig. 2.2 Single-leader-dual-follower teleoperation system. DFR is under the leader robot's control and its movement is unknown to AFR

As X_l is directly from the leader, the desired position for DFR is

$$X_{2,d} = sRX_l, \tag{2.1}$$

where $X_{2,d} = [x_{2,d}, y_{2,d}, z_{2,d}]^T$, R is the rotation matrix which transforms the leader robot's coordinate to the tele-robots' coordinate and s is a scaling factor to be selected according to different implementation scenarios.

Since we focus on the controller design for AFR, its dynamics equation is given as

$$M(q)\ddot{q} + C(q, \dot{q})\dot{q} + G(q) = \tau + J(q)^T F, \tag{2.2}$$

where $q, \dot{q}$ and $\ddot{q}$ are respectively the joint position, velocity, and acceleration of AFR; $M(q)$, $C(q, \dot{q})$, $G(q)$ and $J(q)$ represent the inertial matrix, Coriolis and centrifugal matrix, gravity vector and Jacobian matrix of AFR, respectively; τ denotes the control torque for each joint; $F = [f_x, f_y, f_z]^T$ stands for the force between AFR and the object and f_x, f_y and f_z are components of F in x, y z directions. Within the discussion of this section, the point contact between the robots and the object is considered, i.e., F in (2.3) contains no torques.

Theorem 2.1 *The object held by two robots is assumed to be a spring connecting two EEs and its mass is supposed to be small enough to be neglectable, or is considered estimable [20] such that its effect can be compensated for. Thus, the following development leaves out the object's mass for brevity.*

Based on Theorem 2.1, in accordance with the dynamics of spring, we have

$$||F|| = k_o||\Delta D||, \tag{2.3}$$

where $||F||$ is the holding force of the spring; $\Delta D = [\Delta D_x, \Delta D_y, \Delta D_z]^T$ with ΔD_x, ΔD_y and ΔD_z standing for the increment components between two EEs' displacement in the x, y and z directions; $||\Delta D||$ represents the deformation of the virtual spring; k_o denotes the Hooke's coefficient of the spring; and "$||\cdot||$" stands for 2-norm operator. Then, it is easy to obtain

$$F = k_o \Delta D. \tag{2.4}$$

As shown in Fig. 2.2, the increment is

$$\Delta D = D - (X_1 - X_2), \tag{2.5}$$

where $X_1 = [x_1, y_1, z_1]^T$ and $X_2 = [x_2, y_2, z_2]^T$ represent AFR and DFR's positions respectively and $D = [D_x, D_y, D_z]^T$ stands for the distance between two EEs when the spring is at its rest position, i.e., not deformed. X_1 and X_2 in (2.5) are coordinates in AFR's frame and X_2 is an unknown and time-varying variable. Now, we take the object and DFR as a whole, which is a dynamical environment from AFR's perspective. We define this environment's rest position as X_c (when the object is not deformed), then the contact force becomes

$$F = k_o(X_c - X_1), \quad (2.6)$$

where

$$X_c = [x_c, y_c, z_c]^T = D + X_2, \quad (2.7)$$

which is time-varying because of DFR's unknown movement.

Remark 2.1 It should be pointed out that ΔD should satisfy a condition $0 < \Delta D_{min} < \Delta D < \Delta D_{max}$, where ΔD_{min} is a minimum value to trigger the excitation for parameters estimation and ΔD_{max} is set for system safety. This condition will certainly become the constraint to X_c: $\Delta D_{min} + X_1 < X_c < \Delta D_{max} + X_1$. It means that X_c must deviate from X_1 but can not move too far away from X_1 at every instant. We technically set bounds for X_c to acquire signal excitations and achieve system safety.

2.2.2 *Problem Statement*

The key objective for the SLDF teleoperation presented in Sect. 2.2.1 is to control AFR to collaborate with its partner, DFR, which is directly under the leader's (operator's) control. As X_2 is time-varying and unknown to AFR, AFR's trajectory should be carefully designed, which is the main problem to be addressed in this section.

The task for AFR and DFR is to cooperatively hold and move a deformable object, where the only information known to AFR is the contact force F. To this end, based on F, we firstly need to regulate AFR's position to maintain holding according to DFR's movement; and secondly, a proper trajectory needs to be generated for AFR such that the object's holding force (the contact force F) converges to a desired value.

To simplify the problems, we point out that the situation that two robots hold a deformable object in a hinge-like manner, where no torque is generated between the object and the two robots, unlike grasping, gripping or clamping an object [21–23]. Therefore, we do not consider rotations but only translations of the object.

2.3 Adaptation Scheme

Corresponding to problems stated in Sect. 2.2.2, the proposed method can be divided into three steps: maintaining the holding position of AFR when the object is being translated; estimating X_c that lumps DFR's movement and the object's deformation; and generating a reference trajectory based on impedance control for AFR such that the object's holding force is maintained around the desired value.

2.3.1 Holding Regulation

As there is no torque in F, the object is not expected to rotate in this section, shown in Fig. 2.3a. When DFR translates, AFR should translate in accordance, e.g., AFR should move upwards in Fig. 2.3b as DFR moves upwards. For this purpose, the *holding line* is defined when the contact force is in the horizontal direction, i.e. along x-axis direction in the case of Fig. 2.3. Therefore, the other components of F (f_y and f_z) should be controlled to zero by adjusting y and z positions via

$$\Delta y_1 = -k_y f_y, \tag{2.8}$$

$$\Delta z_1 = -k_z f_z, \tag{2.9}$$

where Δy_1 and Δz_1 denote desired displacements for AFR in the y and z-directions, $k_y > 0$, $k_z > 0$ and they stand for the user defined adaptation gains.

The adaptation scheme in (2.8) and (2.9) makes AFR move against the directions of f_y and f_z until Δy_1 and $\Delta z_1 \to 0$ when f_y and $f_z \to 0$, e.g., AFR should move upwards as f_y points downwards in Fig. 2.3b, which is under a condition that the object has already been compressed. Consequently, AFR will not move away from the holding line and thus the holding line can be maintained along the x axis. Roughly speaking, the adaptation scheme in (2.8) and (2.9) can be interpreted as a special case of feedback force controllers $\Delta y_1 = -k_y(f_y - f_{y,d})$ and $\Delta z_1 = -k_z(f_z - f_{z,d})$, where desired forces $f_{y,d} = 0$ and $f_{z,d} = 0$.

Remark 2.2 Since we have handled issues regarding y_1 and z_1, i.e., AFR's movement is regulated to stay on the holding line. In the following development, we focus on the regulation of x_1, related to x_c and f_x, which are one-dimensional variables.

2.3.2 Unknown Movement Estimation

In this section, we will use the force f_x to estimate the unknown time-varying position x_c, which is critical for generating $x_{1,d}$ in Sect. 2.3.3. Generally, we can model x_c to be a time-varying trajectory as

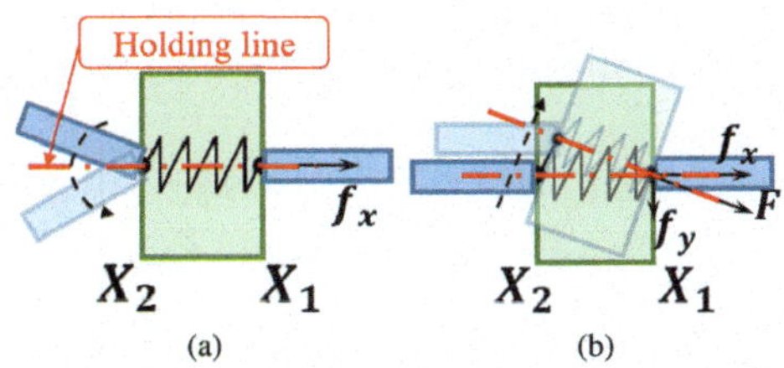

Fig. 2.3 Contact force on AFR changes depending on DFR's movement. **a** No changes on F as DFR rotates. **b** F changes as DFR moves

$$x_c(t) = a_0 + a_1 t + a_2 t^2 + \cdots + a_{n-1} t^{n-1}. \tag{2.10}$$

where t represents time and a_{n-1} with $n = 1, 2, 3, \cdots$ are considered to be constants during a certain period of time. Thus, the corresponding velocity and acceleration are

$$\begin{aligned} \dot{x}_c(t) &= a_1 + 2a_2 t + \cdots + (n-1)a_{n-1} t^{n-2}, \\ \ddot{x}_c(t) &= 2a_2 + \cdots + (n-1)(n-2)a_{n-1} t^{n-3}. \end{aligned} \tag{2.11}$$

where, without loss of generality, we take $n = 3$ in the rest of the discussion for simplicity of analysis.

Based on the model in (2.10), the estimate of x_c is described as

$$\begin{aligned} \hat{x}_c &= \hat{a}_0 + \hat{a}_1 t + \hat{a}_2 t^2, \\ \dot{\hat{x}}_c &= \hat{a}_1 + 2\hat{a}_2 t, \\ \ddot{\hat{x}}_c &= 2\hat{a}_2, \end{aligned} \tag{2.12}$$

where $\hat{*}$ is the estimate of $*$. Then, we define

$$\hat{f}_x = \hat{k}_o(\hat{x}_c - x_1), \tag{2.13}$$

where $\hat{f}_x$ and $\hat{k}_o$ are the estimated force and the object's Hooke coefficient, respectively. Subtracting (2.6) by (2.13), we obtain

$$\begin{aligned} \hat{f}_x - f_x &= \hat{k}_o(\hat{a}_0 + \hat{a}_1 t + \hat{a}_2 t^2 - x_1) \\ &\quad - k_o(a_0 + a_1 t + a_2 t^2 - x_1) \\ &= [x_1 \;\; 1 \;\; t \;\; t^2]\Phi, \end{aligned} \tag{2.14}$$

where

$$\Phi = \begin{bmatrix} -\hat{k}_o + k_o \\ \hat{k}_o \hat{a}_0 - k_o a_0 \\ \hat{k}_o \hat{a}_1 - k_o a_1 \\ \hat{k}_o \hat{a}_2 - k_o a_2 \end{bmatrix}, \tag{2.15}$$

which denotes the estimation errors of the unknown lumped dynamical parameters.

According to Lyapunov theory, we design an adaptive updating law for Φ,

$$\dot{\Phi} = \begin{bmatrix} -\dot{\hat{k}}_o \\ \dot{\hat{k}}_o \hat{a}_0 + \hat{k}_o \dot{\hat{a}}_0 \\ \dot{\hat{k}}_o \hat{a}_1 + \hat{k}_o \dot{\hat{a}}_1 \\ \dot{\hat{k}}_o \hat{a}_2 + \hat{k}_o \dot{\hat{a}}_2 \end{bmatrix} = -\Gamma^{-1} \begin{bmatrix} x_1 \\ 1 \\ t \\ t^2 \end{bmatrix} (\hat{f}_x - f_x), \tag{2.16}$$

i.e., adaptive updating laws for $\hat{k}_o$ and $\hat{a}_i$,

$$\dot{\hat{k}}_o = \gamma_1^{-1} x_1 (\hat{f}_x - f_x), \tag{2.17}$$

$$\dot{\hat{a}}_{j-1} = -\frac{1}{\hat{k}_o}(\gamma_{j+1}^{-1} t^{j-1} (\hat{f}_x - f_x) + \dot{\hat{k}}_o \hat{a}_{j-1}),\ j = 1, 2, 3 \tag{2.18}$$

where the updating rate γ_j in (2.17) and (2.18) belongs to the positive definite matrix $\Gamma = \text{diag}[\gamma_1, \gamma_2, \gamma_3, \gamma_4]$ in (2.16). In the following equations, we show that the adaptation laws in (2.15)-(2.17) make the force estimation error $\hat{f}_x - f_x$ converge to 0.

Let us construct a Lyapunov function as

$$V = \frac{1}{2}\Phi^T \Gamma \Phi. \tag{2.19}$$

Differentiating (2.19), with respect to time, yields

$$\dot{V} = \Phi^T \Gamma \dot{\Phi}. \tag{2.20}$$

Taking the transpose of both sides of (2.14), we have

$$(\hat{f}_x - f_x)^T = \Phi^T \begin{bmatrix} x_1 \\ 1 \\ t \\ t^2 \end{bmatrix}. \tag{2.21}$$

Substituting the adaptive updating law (2.16) into (2.20), and considering (2.21), we have

$$\dot{V} = -\Phi^T \begin{bmatrix} x_1 \\ 1 \\ t \\ t^2 \end{bmatrix} (\hat{f}_x - f_x) = -(\hat{f}_x - f_x)^T (\hat{f}_x - f_x). \tag{2.22}$$

where (2.22) shows that $\dot{V} \leq 0$ and V will decrease when the force estimation error $\tilde{f}_x = \hat{f}_x - f_x$ exists. Therefore, the updating law in (2.16) will eventually lead to $\tilde{f}_x \to 0$ when $t \to \infty$.

2.3.3 *Reference Trajectory Generation*

With the estimation of the unknown movement of DFR and the object deformation, inspired by [16], we develop an algorithm to generate a reference trajectory for AFR, such that the holding force of the held object can be sustained around a fixed value when being moved.

As shown in Fig. 2.2, according to (2.6), we set a desired force

$$f_{x,d} = k_o(x_c - x_{1,d}), \tag{2.23}$$

which corresponds to a desired trajectory

$$x_{1,d} = x_c - \frac{1}{k_o} f_{x,d}, \tag{2.24}$$

where $f_{x,d}$ represents the desired holding force and $x_{1,d}$ denotes the desired trajectory for AFR. Based on Sect. 2.3.2, if we have the estimation of x_c and k_o, we can obtain the desired trajectory for AFR via (2.24), which will be used for comparison in Sect. 2.4 to the forthcoming-developed impedance-model-based approach.

We define a target impedance model for AFR,

$$m\ddot{x}_1 + b\dot{x}_1 + k(x_1 - x_{1,d}) = e_f, \tag{2.25}$$

where

$$e_f = f_x - f_{x,d} \tag{2.26}$$

is the force tracking error, with $f_{x,d}$ as the desired force, m, b and k are positive and represent impedance parameters.

According to (2.6), we can write

$$\begin{aligned} x_1 &= -\frac{1}{k_o} f_x + x_c \\ &= -\frac{1}{k_o}(f_{x,d} + e_f) + x_c. \end{aligned} \tag{2.27}$$

Substituting (2.27) into impedance model (2.25), we obtain

$$\begin{aligned} &m\ddot{e}_f + b\dot{e}_f + (k + k_o)e_f = \\ &-(m\ddot{f}_{x,d} + b\dot{f}_{x,d} + kf_{x,d}) + k_o(m\ddot{x}_c + b\dot{x}_c + kx_c) - kk_o x_{1,d}. \end{aligned} \tag{2.28}$$

As $f_{x,d}$ is constant, we have

$$\begin{aligned} &m\ddot{e}_f + b\dot{e}_f + (k + k_o)e_f = \\ &-kf_{x,d} + k_o(m\ddot{x}_c + b\dot{x}_c + kx_c) - kk_o x_{1,d}. \end{aligned} \tag{2.29}$$

For such force tracking error dynamics, the steady-state error is

$$e_{f,ss} = \frac{k_o}{k + k_o}\left(-\frac{k}{k_o} f_{x,d} + (m\ddot{x}_c + b\dot{x}_c + kx_c) - kx_{1,d}\right). \tag{2.30}$$

Consequently, to eliminate the steady-state error, we need the desired AFR trajectory to satisfy

$$x_{1,d} = \frac{1}{k}(m\ddot{x}_c + b\dot{x}_c + kx_c - \frac{k}{k_o} f_{x,d}). \tag{2.31}$$

Replacing x_c and k_o in (2.31) with the updating $\hat{x}_c$ and $\hat{k}_o$ from (2.23), we will have AFR's reference trajectory

$$x_{1,r} = \frac{1}{k}(m\ddot{\hat{x}}_c + b\dot{\hat{x}}_c + k\hat{x}_c - \frac{k}{\hat{k}_o} f_{x,d}). \tag{2.32}$$

As similarly proved in [16], we eventually have $f_x \to f_{x,d}$ as $x_{1,r}$ eliminates $e_{f,ss}$.

To summarise, we have $\hat{f}_x \to f_x \to f_{x,d} \to ||F_d||$ because i) $\hat{f}_x \to f_x$ as described in the previous subsection, ii) $f_x \to ||F||$ as f_y and $f_z \to 0$ and iii) $f_{x,d}$ is set as $||F_d||$ (e.g., $F_d = [f_{x,d}, 0, 0]^T$ in this development). This means that based on the holding line regulation, the movement-estimation-based and impedance-model-based reference trajectory $x_{1,r}$ leads to the object's contact force converging to the desired value.

2.4 Simulations and Experimental Study

Since the adaptation of AFR's reference trajectory requires a certain time to stabilize, we smooth it via $x'_{1,r} = x_{1,r}(1 - e^{-\mu t^\eta})$. In this way, $x'_{1,r}$ will start from zero and will be gradually dominated by $x_{1,r}$ as time goes by. $\mu = 5$ and $\eta = 1$ in all simulations and $\mu = 0.01$ and $\eta = 2$ in all experiments.

2.4.1 *Simulations*

Three groups of simulations are conducted on two 3-DOFs robots in MATLAB, as shown in Fig. 2.4a, whose task space is on a two-dimensional plane. The two robots' dynamic and kinematic parameters are presented as follows. Robot links' lengths are l_{i1}, l_{i2}, l_{i3}=0.5, 0.3, 0.1 (m). Their masses are m_{i1}, m_{i2}, m_{i3}=1.5, 0.9, 0.2 (kg). AFR base coordinate is [0.5, 0] (m). DFR base coordinate is [0, 0] (m). There are two initial positions of AFR, which are shown in Fig. 2.4a: $[1/3\pi, 1/3\pi, 1/3\pi]^T$(rad) and in Fig. 2.4b: $[0.9112, 1.4184, 0.8120]^T$(rad), and they will be utilized alternatively for comparison purposes. DFR initial joint position is $[2/3\pi, -1/3\pi, -1/3\pi]^T$(rad). PD controller gains are Kp=diag[700, 600, 500], Kd=diag[10, 1, 0.1]. Sampling time is 10^{-3} (s) and scaling factor $s = 1$. The holding line direction is parallel to the x-axis.

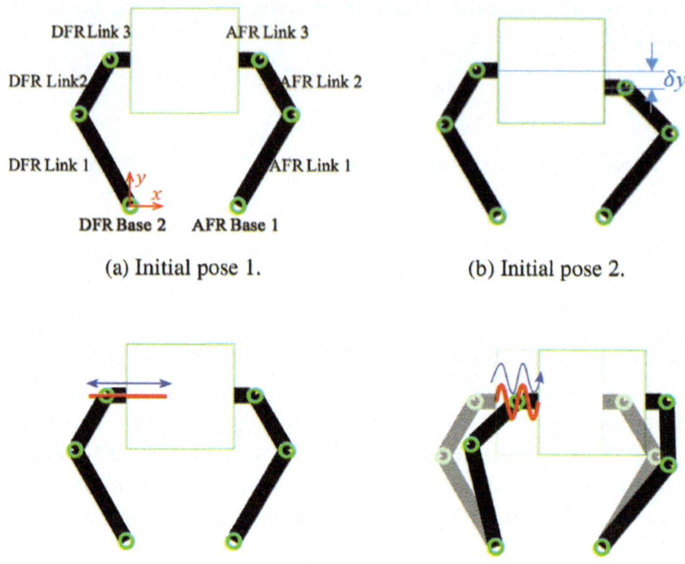

(a) Initial pose 1. (b) Initial pose 2.

(c) Traj. 1. (d) Traj. 2.

Fig. 2.4 The simulation platform with two 3-DOFs robots and simulation settings. **a, b** are two initial poses of the two robots, alternatively used in the simulations. In **c, d**, the red lines represent DFR's trajectories and the blue lines with arrows indicate the moving directions of the trajectories

Two trajectories for X_l ($X_{2,d}$) illustrated in Fig. 2.4c, d are set to mimic the operator's movement. Traj. 1: $\{(x_l, y_l)|x_l = \alpha \sin \frac{2\pi}{\sigma} t, y_l = 0\ \}$. Traj. 2: $\{(x_l, y_l)|x_l = \beta t, y_l = \alpha \sin \frac{2\pi}{\sigma} t\ \}$. They are also designed for comparison purposes and will be implemented alternatively in the forthcoming simulations, where α, β and σ will be adjusted as needed.

2.4.1.1 Holding Line Regulation

In this group of simulations, we consider the following two cases:
Case 1: Due to calibration error or measurement noise, AFR and DFR may start a task from two different holding lines. In this part, we simulate such a case by commanding the two robots to hold and move an object with DFR teleoperated by Traj. 1 and two robots' initial positions in Fig. 2.4b. To address the effect of the misalignment, Eq. (2.8) is applied with $k_y = 0.001$. In this case, we set $k_o = 45$, $f_{x,d} = 1$N, and in Traj. 1, $\alpha = 0.18$ and $\sigma = 20$. x_2 and k_o are assumed to be known to AFR.

The holding force $||F||$ is presented in Fig. 2.5a. In this case, when (2.8) is not applied, $||D||$ is larger due to the discrepancy between two robots' actual moving directions, compared to the situation where (2.8) is utilized. As a result, $x_{1,r}$ also gets larger, leading to a larger distance between two EEs, which contributes to a larger error in force tracking.

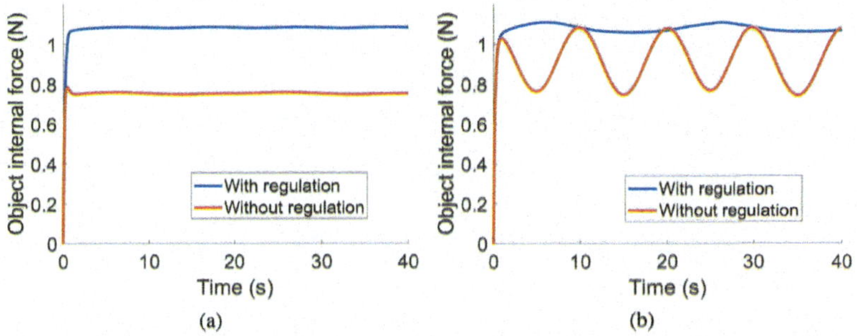

Fig. 2.5 The holding forces in Case 1 and Case 2 in Sect. 2.4.1.1

Case 2: Even if both EEs are perfectly aligned in a holding line, when there is environmental disturbance acting on DFR, DFR's movement will deviate from its desired direction, which results in δy between the two EEs. In this part, we mimic such a case by using Traj. 2 and Fig. 2.4a, where $\beta = 1/200$ and other settings are the same as in Case 1. Results are illustrated in Fig. 2.5b.

From the above two cases, we can find how the regulation in (2.8) enables AFR to move up and down such that the holding line is maintained efficiently. In addition, its significance is evident in the case when DFR's deviation is relatively large. Such a case may cause failure in the task of holding and moving an object due to the insufficiently large holding force.

2.4.1.2 Estimation of Unknown Movement

In this group of simulations, using the proposed estimation approach in Sect. 2.3.2, we will demonstrate that the desired contact force can be achieved even if we set parameters k_o and x_1 to different values. Two cases are involved, where x_c is unknown to AFR and initial values of the variables are: $\hat{k}_o = 30$ (N/m) (unit "N/m" are the same for all stiffness variables in this section and will be left out in the following), $\dot{\hat{k}}_o = 0$, $\hat{a}_0 = 0.55$, $\dot{\hat{a}}_0 = 0$, $\hat{a}_1 = 0$, $\dot{\hat{a}}_1 = 0$, $\hat{a}_2 = 0$, $\dot{\hat{a}}_2 = 0$, $m = 0.01$, $b = 0.1$ and $k = 10$. The initial position in Fig. 2.4a and Traj. 1 are employed.
Case 1: In this case, we set $k_o = 25$, 35 and 45 respectively, $\alpha = 0.18$ and $\sigma = 20$. Other parameters are the same as those in Sect. 2.4.1.1 Case 1.

As shown in Fig. 2.6, f_x converges to $f_{x,d}$ in all three conditions.
Case 2: As the movement X_2 of DFR, is unknown to AFR, we will demonstrate that when the desired force is set close to 0, the proposed adaptive estimation method can make $\hat{x}_c$ converge to x_c despite different x_2. Various x_2 are set by defining different σ in Traj. 1: σ=10, 30 and 40. k_o is set to 35 in this case and other parameters are the same as those in Sect. 2.4.1.1 Case 1. Simulation results are illustrated in Fig. 2.7.

As shown in Fig. 2.7, when DFR's movement x_2 changes with σ varying from 10 to 40, $\hat{x}_c$ eventually converges to x_c (the first row) with estimation errors converging

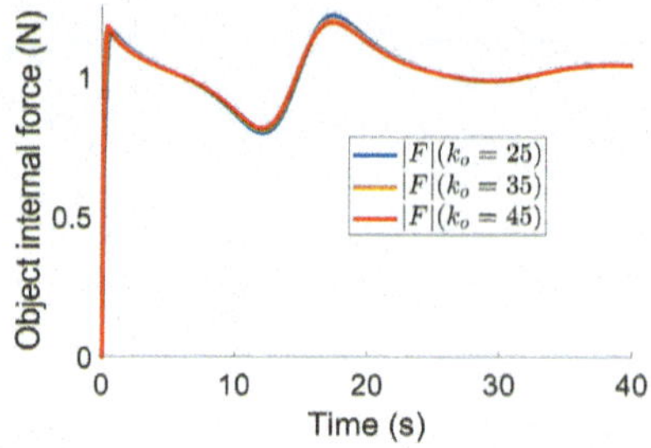

Fig. 2.6 Results of Case 1 in Sect. 2.4.1.2

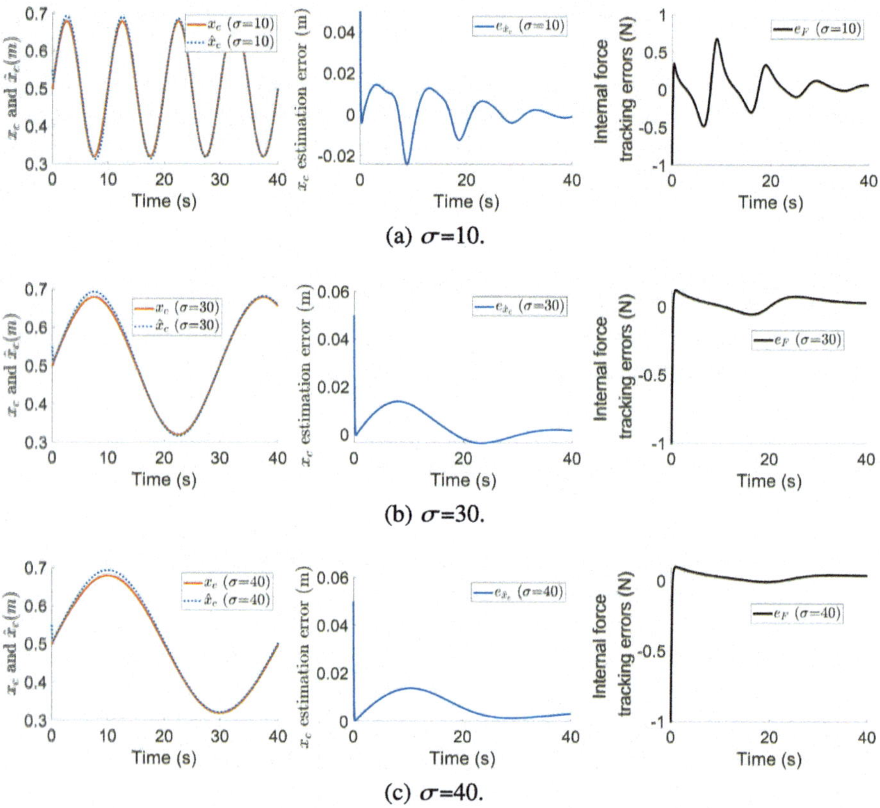

Fig. 2.7 Results of Case 2 in Sect. 2.4.1.2

to 0 (the middle row), although the convergence takes a longer time when x_2 changes fast ($\sigma = 10$). Such an estimation makes the object's holding force converge to $f_{x,d}$ ($||F_d||$), as its tracking errors fall into a small range around 0, depicted in the third row of Fig. 2.7.

2.4.1.3 Comparison with Other Methods

To exhibit the superiority of our proposed method in detail, we design the following comparative simulations, where the proposed method in (2.32) is compared to the conventional force controller, the passive impedance controller [15] and the method described in (2.24). The force controller is described as $x_1 = 0.02(f_x - f_{x,d})$. The passive impedance controller is designed based on $e_f = m_{pi}\ddot{x}_1 + b_{pi}\dot{x}_1$ and is solved via the second-order Runge-Kutta method, where $m_{pi} = 3$ and $b_{pi} = 2$.

In this comparison, we take both Traj. 1 and 2 and the robots' initial position of Fig. 2.4b into consideration, simulating a situation close to reality. $k_o = 35$ and other parameters are the same as those in Sect. 2.4.1.2. Eighteen simulations under different configurations are displayed in Table 2.1, where "F" represents the conventional force controller, "PIm" stands for passive impedance model-based controller, "NoIm" is the non-impedance-based controller (2.24) and "Im" denotes our proposed impedance-based controller (2.32).

We evaluate these methods by comparing the holding force tracking errors. As shown in Fig. 2.8, we deliberately divide eighteen simulations into three groups according to differences between their tracking trajectories and f_d. Group 1 is in the light grey table. They track Traj. 1 and $f_d = 1$ and the corresponding results are defigsted in Fig. 2.8a. Group 2 is in the medium grey table. They track Traj. 2 and $f_d = 1$ and the corresponding results are depicted in Fig. 2.8b. Group 3 is in the dark grey table. They track Traj. 2 and $f_d = 7$ and the corresponding results are depicted in Fig. 2.8c. Figure 2.8a demonstrates that the force controller achieves the worst performance, so it is not considered in the rest of the simulations. Comparing

Table 2.1 Please write your table caption here

No.	s1	s2	s3	s4	s5	s6	s7	s8
Controller	F	F	PIm	PIm	NoIm	NoIm	Im	Im
Traj. No.	1	1	1	1	1	1	1	1
Holding line regulation	No	Yes	No	Yes	No	Yes	No	Yes
f_d (N)	1	1	1	1	1	1	1	1
No.	s9	s10	s11	s12	s13	s14		
Controller	PIm	PIm	NoIm	NoIm	Im	Im		
Traj. No.	2	2	2	2	2	2		
Holding line regulation	No	Yes	No	Yes	No	Yes		
f_d (N)	1	1	1	1	1	1		
No.	s15	s16	s17	s18				
Controller	NoIm	NoIm	Im	Im				
Traj. No.	2	2	2	2				
Holding line regulation	No	Yes	No	Yes				
f_d (N)	7	7	7	7				

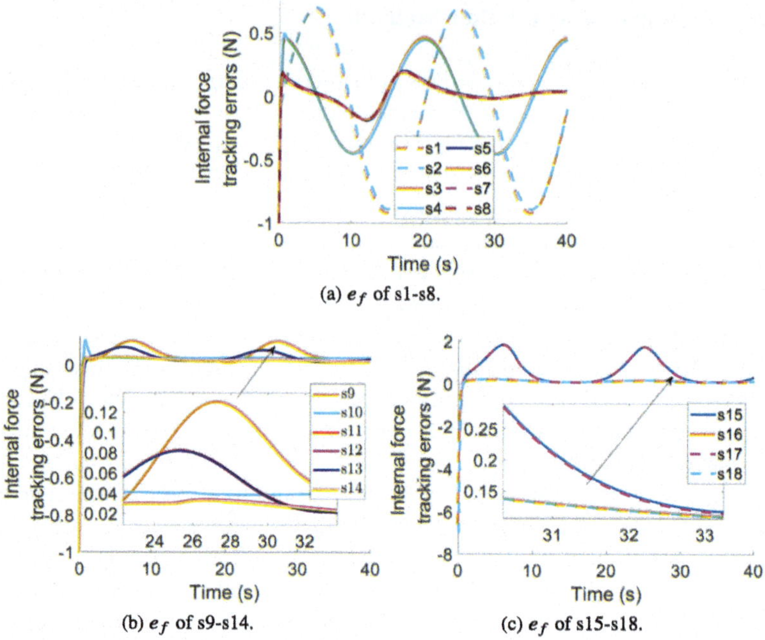

(a) e_f of s1-s8.

(b) e_f of s9-s14.

(c) e_f of s15-s18.

Fig. 2.8 Holding force tracking results of s1–s18

PIm to NoIm and Im in Fig. 2.8a and Fig. 2.8c, it is obvious that NoIm and Im have better performance than PI. Comparing results of s3 to s4, s5 to s6, ..., and s17 to s18, where the difference between them is whether (2.8) is applied or not, we can conclude that (2.8) plays a significant role in improving the regulation of the holding line and further maintaining the holding force. From Figs. 2.8b, c, we can tell that as the desired force becomes larger, the superiority of (2.8) becomes more evident. Similarly, comparing s5 and s6 to s7 and s8, s11 and s12 to s13 and s14, s15 and s16 to s17 and s18, we find that they all perform well because they all implement the movement estimation. Moreover, we can conclude that Im performs better than NoIm because, except for the movement estimation, Eq. (2.32) takes $\ddot{\hat{x}}_c$ and $\dot{\hat{x}}_c$ as parts of the control input, which has the ability to predict the movement of x and adjust $x_{1,r}$.

2.4.2 Experiments

We design a task for the SLDF teleoperation system to hold and move an object on a physical robot platform. The experiment setup is shown in Fig. 2.9a, where the TouchX device is regarded as the single leader device and is manipulated by the operator, two arms of Baxter robot are treated as the dual tele-robots, a force sensor

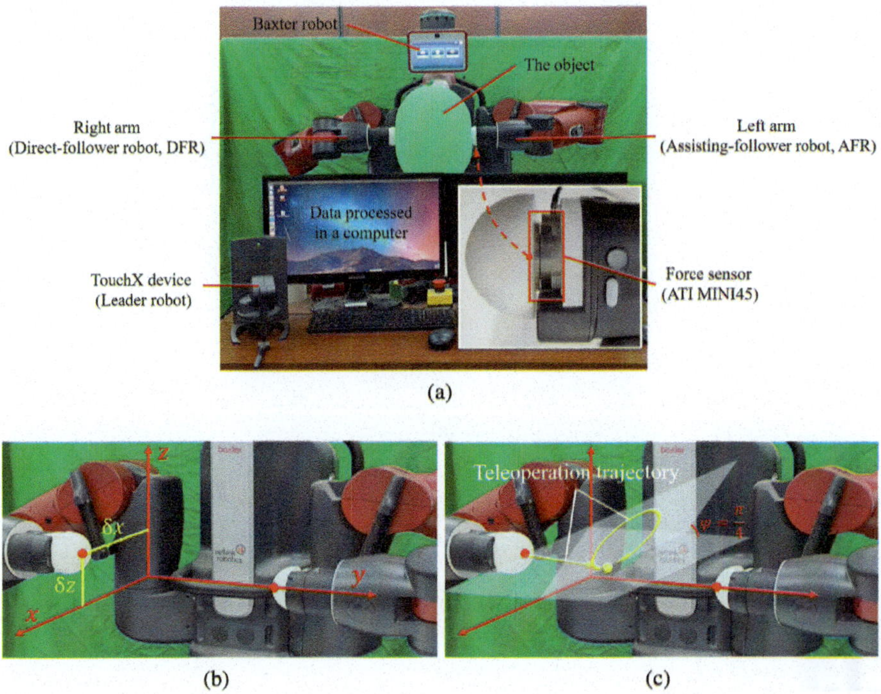

Fig. 2.9 The experimental setup. **a** The two arms of the Baxter robot are viewed as two independent follower robots, i.e., AFR's motion regulation is generated not based on any information of DFR. **b** Discrepancies δx and δz while y direction is chosen as the holding line direction. **c** The designed teleoperation trajectory contains movements in three directions

(ATI MINI45 F/T) is mounted on the left arm EE and a balloon is considered as the object to be held and moved. Baxter's right arm is directly teleoperated by the TouchX and the movement of the left arm is independently controlled based on the contact force but without using any information of the right arm. To demonstrate the holding line regulating ability, the initial poses of two EEs are set with discrepancies in x and z directions, while y axis direction is chosen as the holding line, as shown in Fig. 2.9b. We set $s = 1.2$, $R = [0, -1, 0; 1, 0, 0; 0, 0, 1]$ and $n = 2$. We conduct the following two experiments to demonstrate how the proposed method works.

2.4.2.1 Holding Regulation in Static Situation

We keep DFR static, while AFR is controlled to approach the object along the y axis. When AFR establishes its contact with the object with a force above 1N, the proposed method is triggered in order to eliminate the discrepancies in x and z directions between two robots' EEs and to maintain the desired holding force. We set two discrepancies: $[\Delta x, \Delta z] = [0.08, 0.08]$m and $[0.1, 0.1]$m. Parameters are:

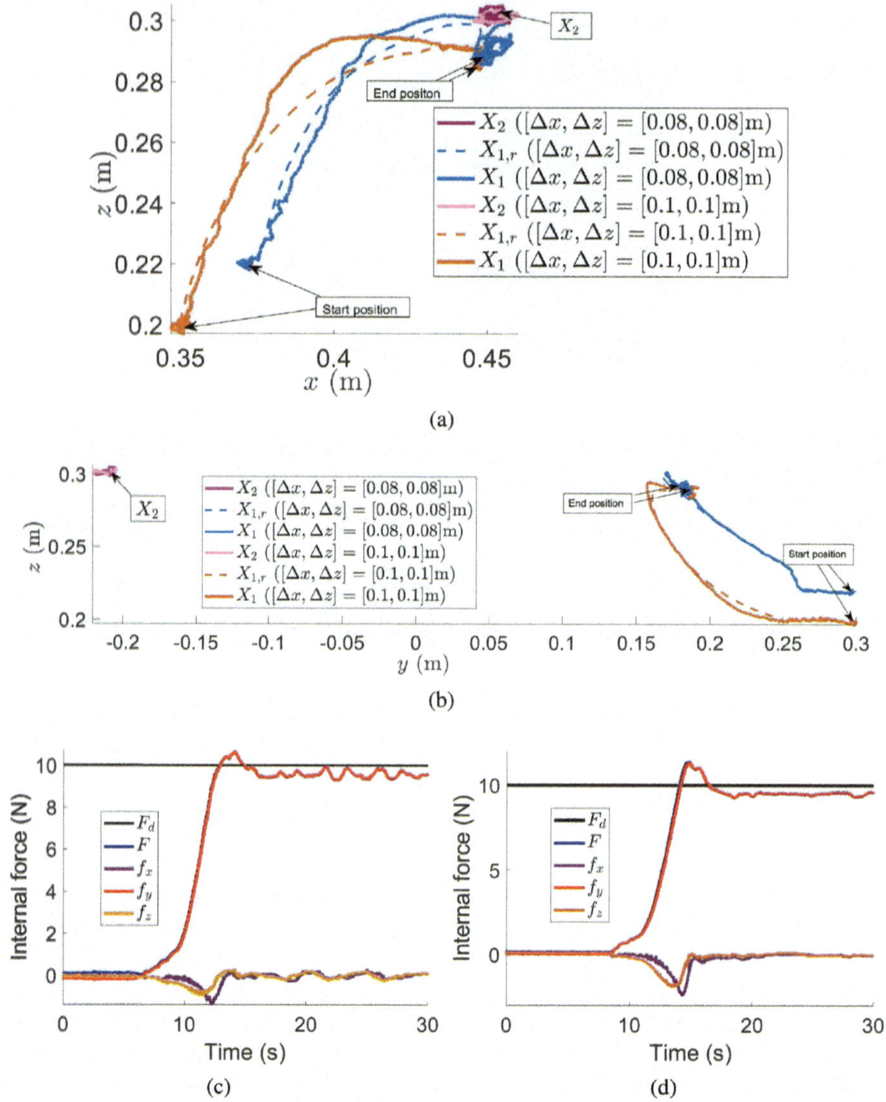

Fig. 2.10 **a** EE trajectories projection in xoz-plane. **b** EE trajectories projection in yoz-plane. **c** Forces with initial discrepancy $[\Delta x, \Delta z] = [0.1, 0.1]$ m. **d** Forces with initial discrepancy $[\Delta x, \Delta z] = [0.08, 0.08]$ m

$k_x = k_z = 0.0002$, initial value $\hat{k}_o = 10, \hat{a}_0 = 0$, $f_{y,d} = 10$ N and others are the same as those in Sect. 2.4.1.2. The experimental results are shown in Fig. 2.10.

From Fig. 2.10a, we can tell that AFR adjusts its position to hold the object properly with the proposed holding line regulating method, no matter where its initial position is. From Fig. 2.10b, we see AFR moves forward and when it contacts the object (where $y \approx 0.25$m), the movement-estimation algorithm is triggered to adjust AFR and finally move it to the same destination as they are required to make the holding forces converge to the same F_d depicted in Fig. 2.10c, d. Figure 2.10c, d also demonstrate that f_x and f_z converge to zero and f_y converges to $f_{y,d}$.

2.4.2.2 Continuous Movement

DFR is teleoperated along a designed trajectory shown in Fig. 2.9c, which stands for a general teleoperation task that has combinations of movements in x, y and z directions, including a straight line and a curve. The initial discrepancy is set to $[\Delta x, \Delta z] = [0.08, 0.08]$m and the other parameters are the same as those in the previous experiment. For comparison purposes, the force controller and the passive impedance controller described in the simulations are also tested. The experimental results are shown in Fig. 2.11.

From Fig. 2.11a, b, we can see that after AFR adjusts its reference trajectory to stay on the holding line, the two EEs have an approximately fixed distance, which leads to the object's holding force F staying around the desired value. From Fig. 2.11c, we can find that the proposed method has better performance compared with the conventional force controller and the passive impedance controller in terms of force tracking. This is more obvious after about 20 s, corresponding to the circle in Fig. 2.9c which includes forward and backward movements. A similar result can be seen in Fig. 2.8a, where if DFR moves forwards and backward in the holding line direction, the advantage of the proposed method in regulating the holding force becomes more evident. Therefore, we can conclude that (i) the proposed method is effective in regulating the holding line and maintaining the contact force; and (ii) the movement estimation in the proposed method is important in dealing with time-varying movements.

2.5 Conclusion

This chapter introduces a method for an SLDF teleoperation system to hold and move an object, with the object's holding force maintained around a desired value. Three necessary steps are included. Firstly, the holding line regulation method is to make the robots' holding force orientation stay the same. Secondly, an adaptive movement estimation approach is used to enable the assisting tele-robot to infer its partner's movement. Thirdly, an impedance controller is derived such that the desired contact force is achieved. Comparative simulations and experiments are designed and their results verify the feasibility and effectiveness of the proposed method.

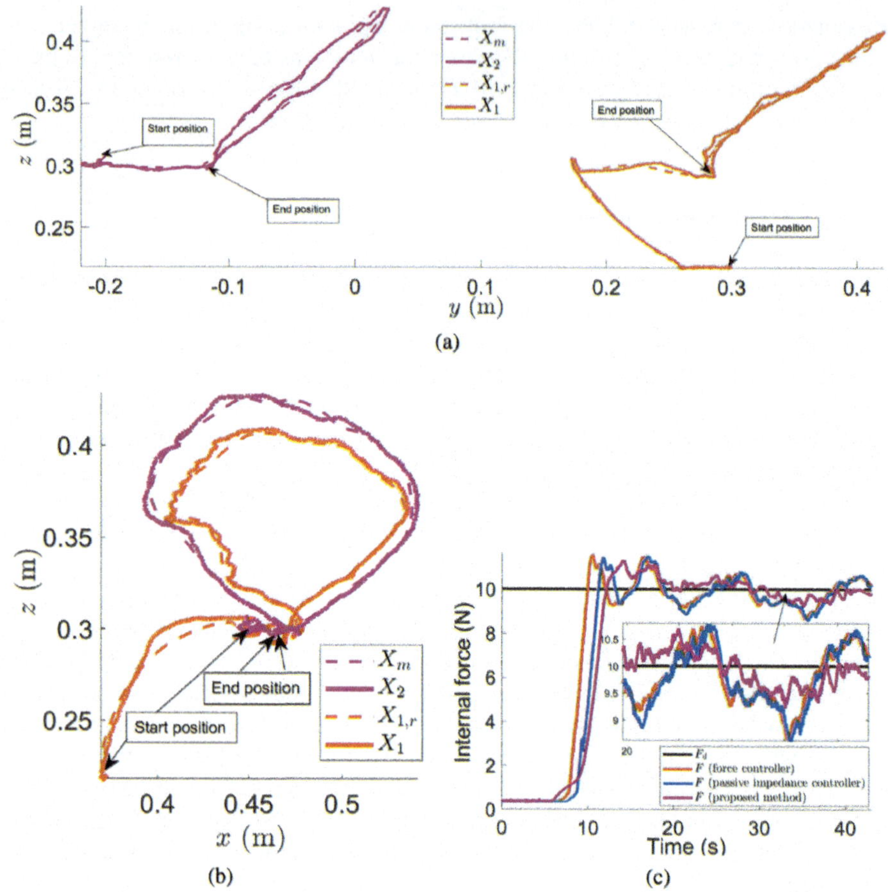

Fig. 2.11 **a** EE trajectories projection in yoz-plane. **b** EE trajectories projection in xoz-plane. **c** Object's holding forces under three methods

References

1. Li Z, Su C (2013) Neural-adaptive control of single-master-multiple-slaves teleoperation for coordinated multiple mobile manipulators with time-varying communication delays and input uncertainties. IEEE Trans Neural Netw Learn Syst 24(9):1400–1413
2. Cheung Y, Chung JH et al (2009) Semi-autonomous formation control of a single-master multi-slave teleoperation system. In: Paper presented at the 2009 IEEE symposium on computational intelligence in control and automation, Nashville, TN, USA, 30 Mar 2009–02 Apr 2009
3. Sun D, Liao Q et al (2020) Single master bimanual teleoperation system with efficient regulation. IEEE Trans Robot 36(4):1022–1037
4. Sun Y, Pan B et al (2021) Vision-based framework of single master dual slave semi-autonomous surgical robot system. IRBM 42(1):55–64
5. Watanabe K, Kanno T et al (2017) Single-master dual-slave surgical robot with automated relay of suture needle. IEEE Trans Indus Electron 65(8):6343–6351

6. Frank B, Schmedding R et al (2010) Learning the elasticity parameters of deformable objects with a manipulation robot. In: Paper presented at the 2010 IEEE/RSJ international conference on intelligent robots and systems, Taipei, Taiwan, 18–22 Oct 2010
7. Sanchez J, Corrales J et al (2018) Robotic manipulation and sensing of deformable objects in domestic and industrial applications: a survey. Int J Robot Res 37(7):688–716
8. Yin H, Varava A et al (2021) Modeling, learning, perception, and control methods for deformable object manipulation. Sci Robot 6(54):eabd8803
9. Liang W, Feng Z et al (2021) Dexterous manoeuvre through touch in a cluttered scene. In: Paper presented at the 2021 IEEE international conference on robotics and automation (ICRA), Xi'an, China, 30 May 2021–05 Jun 2021
10. Liang W, Feng Z et al (2020) Robust force tracking impedance control of an ultrasonic motor-actuated end-effector in a soft environment. In: Paper presented at the 2020 IEEE/RSJ international conference on intelligent robots and systems (IROS), Las Vegas, NV, USA, 24 Oct 2020–24 Jan 2021
11. Li D, Yu H et al (2021) On time-synchronized stability and control. IEEE Trans Syst Man Cybern Syst 52(4):2450–2463
12. Hirata Y, Wang Z et al (2009) Transporting an object by a passive mobile robot with servo brakes in cooperation with a human. Adv Robot 23(4):387–404
13. Iqbal K, Zheng YF (1999) Arm-manipulator coordination for load sharing using predictive control. In: Paper presented at the proceedings 1999 IEEE international conference on robotics and automation (Cat. No.99CH36288C), Detroit, MI, USA, 10–15 May 1999
14. Iqbal K, Inooka H (1995) Variable impedance control of a robot for cooperation with a human. In: Paper presented at the proceedings of 1995 IEEE international conference on robotics and automation, Nagoya, Japan, 21-27 May 1995
15. Ficuciello F, Villani L et al (2015) Variable impedance control of redundant manipulators for intuitive human-robot physical interaction. IEEE Trans Robot 31(4):850–863
16. Seraji H, Colbaugh R (1997) Force tracking in impedance control. Int J Robot Res 16(1):97–117
17. Jung S, Hsia TC et al (2004) Force tracking impedance control of robot manipulators under unknown environment. IEEE Trans Control Syst Technol 12(3):474–483
18. Zhang X, Khamesee MB (2017) Adaptive force tracking control of a magnetically navigated microrobot in uncertain environment. IEEE/ASME Trans Mechatron 22(4):1644–1651
19. Stephens TK, Awasthi C et al (2019) Adaptive impedance control with setpoint force tracking for unknown soft environment interactions. In: Paper presented at the 2019 IEEE 58th conference on decision and control (CDC), Nice, France, 11–13 Dec 2019
20. Ćehajić D, Hirche S et al (2017) Estimating unknown object dynamics in human-robot manipulation tasks. In: Paper presented at the 2017 IEEE international conference on robotics and automation (ICRA), Singapore, 29 May 2017–3 Jun 2017
21. Wojtara T, Uchihara M et al (2009) Human-robot collaboration in precise positioning of a three-dimensional object. Automatica 45(2):333–342
22. Lee SY, Lee KY et al (2007) Human-robot cooperation control for installing heavy construction materials. Autonom Robot 22:305–319
23. Agravante DJ, Cherubini A et al (2014) Collaborative human-humanoid carrying using vision and haptic sensing. In: Paper presented at the 2014 IEEE international conference on robotics and automation (ICRA), Hong Kong, China, 31 May 2014–7 June 2014

Part II
Integrated Autonomous Learning and Control Framework

Chapter 3
A Small Opening Workspace Control Strategy for Redundant Manipulator Based on Remote Center of Movement Method

This chapter presents a method for redundant manipulators working in the small opening workspace without collision. To achieve this aim, we began with an improved incremental Radial Basis Function Neyral Network (RBFNN) method to estimate manipulator dynamic parameters, and then with the help of Lynapunov function, the control strategy could converge within a fixed-time. To avoid the collision of workspace and constrain the posture of end-effector, we proposed a safety region CNN method adapted with the Remote Center of Motion method inspired by the minimally invasive surgical manipulator. Torque observer is also implied to estimate the external force to resist external interference. Experiments on Baxter, a 7-DoF redundant manipulator, demonstrate the feasibility of the proposed control strategy.

3.1 Introduction

In recent years, with the continuous research on neural networks, researchers have made remarkable progress in manipulator control with uncertain kinematics parameters or dynamics model [1–3]. Among all, the RBFNN method has been proved that RBFNN is a reliable way of estimating unknown parameters of manipulator. As shown in [4, 5], RBFNN was applied to the parameter estimation of the robot with uncertain parameters, and RBFNN was trained based on the trajectory tracking error to ensure the stability of the robot system. The Lyapunov function proved that the position tracking error would converge to the specified boundary within the wired time. The simulation results show that the control method based on RBFNN parameter identification has better stability. However, in traditional RBFNN, the parameters of neural nodes need to be designed in advance. In this case, the parameter estimation ability of neural network will be affected if the neural network parameters were not set properly, leading to the deterioration of the control effect of the manipulator.

C. Yang et al., *Advanced Teleoperation and Robot Learning for Dexterous Manipulation*, Springer Tracts in Advanced Robotics 160,
https://doi.org/10.1007/978-3-031-78501-6_3

Researchers proposed incremental learning to improve traditional RBFNN method with adaptive node parameters designed [6].

Neural networks and deep learning have gained great attention in the research of computer vision, where grasping detection, as one of the representative research fields, has produced a lot of research results. Convolutional neural network is one of the most commonly used methods for its good effect and high classification accuracy [7–9]. An accurate and real-time robot grasping detection method based on convolutional neural network was proposed in [7], which could be modified to predict multiple fetches per object by using a local constraint prediction mechanism. In [8] the deep convolution neural network is used to extract features from the scene, and then the shallow convolution neural network is used to predict the grasping configuration of interested objects. In order to integrate the information of various sensors and improve the accuracy of grasping detection, in [10] a hybrid deep structure of robot grasping and detecting based on visual and tactile sensing is proposed.

According to the existing studies, for manipulators operating with window-shaped obstacles, the window-shaped obstacles would usually be regarded as the limitation in the joint space of the manipulator in trajectory planning task [11, 12], or the precise matching of grasping objects and holes in peg-in-hole tasks [13, 14]. However, in [11, 12], the researchers only considered the case where the end of the manipulator slightly passed through the window-shaped obstacle for less than $0.1m$, while in [13, 14], the peg-in-hole tasks, the researchers were more concerned with the precision of the end of the manipulator. In these researches, the end-effector was not considered with the task of operating after passing through the window-shaped obstacle. A similar problem of operating in a workspace with limited openings were found, whose idea came from the minimally invasive surgery as the remote center of movement (RCM) constraint. Su *et al.* proposed the application of RCM constraint on redundant surgical robots in [15], and the experimental results proved that this control strategy had a good control effect. Later in [16] the end effector accuracy and the compliance control performance were improved through the adaptive compensator method. This chapter refers to the RCM constraint control strategy of the minimally invasive surgical robot and integrate it into the control strategy of the industrial manipulator in order to successfully achieve the assembly task in the small opening space.

The key innovations and contributions of this chapter are shown as follows:

1. Innovatively adopted the fixed-time control strategy to the null space control method for redundant manipulator operation to ensure the control effect of the main task and accomplish the sub-task at the same time.
2. Based on the RCM control strategy, the existing control strategy of window-shaped obstacles avoidance is improved, and the end effector of the manipulator can pass through the center of the window-shaped obstacle while successfully grasp the target without collision.

3. A safety boundary control strategy was adopted based on the traditional visual grasping selection based on CNN classifier. Considering the influence of workspace boundaries within grasping tasks, the object grasping control strategy in workspace was optimized to avoid the collision between grasping tasks and workspace boundaries.

3.2 Methodology

3.2.1 Dynamic Setup

For a redundant manipulator with n degrees of freedom (DoF) executing the task in a m dimensional task space, the dynamic model can be represented as:

$$M(q)\ddot{q} + C(q,\dot{q})\dot{q} + G(q) = \tau_C - \tau_{Ex}, \tag{3.1}$$

where vector $q = [q_1, \ldots, q_n]^T \in \mathbb{R}^n$ and vector $G(q) \in \mathbb{R}^n$ represent n joint angles of the redundant manipulator and gravitational moment respectively, while matrix $M(q) \in \mathbb{R}^{n\times n}$ and matrix $C(q,\dot{q}) \in \mathbb{R}^{n\times n}$ represent inertia matrix and the Coriolis and Centrifugal effects. Vectors $\tau_C \in \mathbb{R}^n$ and $\tau_{Ex} \in \mathbb{R}^n$ respectively denote the control torque input and the external torque applied on the manipulator.

Property 3.1 [17] For any $q \in \mathbb{R}^n$ and $\dot{q} \in \mathbb{R}^n$, $M(q)$, $C(q,\dot{q})$ and $G(q)$ are bounded and first-order derivable.

Property 3.2 [17] The equation $\vartheta^T \left(\dot{M} - 2C\right)\vartheta = 0$ holds for any $\vartheta \in \mathbb{R}^n$.

3.2.2 Incremental RBF Neural Networks

It has been proved that the RBFNN has good performance on global approximation. Consider $Y(z) \in \mathbb{R}^n$ as the nonlinear part needs to be approximated, which could be written as:

$$Y(z) = W^T S(z) + \epsilon(z), \tag{3.2}$$

where $z \in \mathbb{R}^m$ denotes the input of the neural network, $W \in \mathbb{R}^{n\times k}$ represents the weighting matrix of the neural network, $S(z) \in \mathbb{R}^{k\times 1}$ represents the regression vector and caused by the limited number of neural nodes, the approximation error is represented with $\epsilon \in \mathbb{R}^n$ assuming is bounded as $0 \leqslant \|\epsilon\| \leqslant \bar{\epsilon}$, $\bar{\epsilon} \in \mathbb{R}^n$.

The following Gaussian function is chosen as the basic function:

$$S_i(z) = exp\left(\frac{-(z-\zeta_i)^T (z-\zeta_i)}{\sigma_i^2}\right), i = 1, \ldots, k, \tag{3.3}$$

where ζ_i and σ_i represents the i-th center point and width of neural network.

Compared with the traditional RBFNN method requires an advance design center node, we adopt the dynamic generation center incremental RBFNN method, based on the input and the status of existing nodes decide the generation of new nodes. The updated law of the center node set is designed as:

$$\zeta\,(t+T)=\begin{cases}[\zeta\,(t)\ \zeta_{new}], & d_{node}>\varepsilon\\ \zeta\,(t), & d_{node}\leqslant\varepsilon\end{cases} \tag{3.4}$$

where d_{node} represents the telorism between the previous node set and the current input, ε represents the threshold to be designed and ζ_{new} represents the new node set. The generation formulas are given as follows:

$$\zeta_{new}=\bar{\zeta}_{min}+\delta\cdot\left(z-\bar{\zeta}_{min}\right) \tag{3.5}$$

$$\bar{\zeta}_{min}=\frac{\sum_{i=1}^{m}\zeta_i}{m} \tag{3.6}$$

$$dis=\left\|-\bar{\zeta}_{min}\right\| \tag{3.7}$$

where z represents the current input, $\delta>0$ is a parameter to designed and $\bar{\zeta}_{min}$ represents the weighted average node of the m center nodes closest to the current input.

3.2.3 Preliminaries

Definition 3.1 [18] For any continuous system $\dot{z}=f\,(z,t)$ with the initial state $z=z_0$ at $t=0$, if there exists a real number T_{max} and the convergence time of the system T satisfies $0\leqslant T\leqslant T_{amx}$, the system could be considered as a fixed-time stable system.

Lemma 3.1 *[19] Define a positive first-order derivable Lyapunov function $V\,(z)$ with a bounded initial state $V\,(z_0)$, and there always exist positive parameters p and q satisfying: $\dot{V}<-pV+q$, then it could be drawn that the system signals are uniformly ultimately bounded(UUB).*

Lemma 3.2 *[20] Define the positive first-order derivable function $V(z)$ as the Lyapunov function of the system $\dot{z}=f(z,t)$ and a compact set:*

$$\Omega=\left\{z\mid V\leqslant\min\left(\left[\frac{h}{m(1-\gamma)}\right]^{1/\alpha},\left[\frac{h}{n(1-\gamma)}\right]^{1/\beta}\right)\right\}.$$

If there always exist $m>0$, $n>0$, $h>0$, $0<\alpha<1$ and $\beta>1$, satisfying

$$\dot{V} < -mV^{\alpha} - nV^{\beta} + h,$$
$$0 \leqslant T \leqslant T_{max} = \frac{1}{m\gamma(1-\alpha)} + \frac{1}{n\gamma(\beta-1)}, \tag{3.8}$$

with $0 < \gamma < 1$.

Lemma 3.3 *[21] The following inequalities holds for any* x_i*:*

$$\sum_{i=1}^{m} |x_i|^k \geqslant \left(\sum_{i=1}^{m} |x_i|\right)^k, \ for\ 0 < k \leqslant 1 \tag{3.9}$$

$$\sum_{i=1}^{m} |x_i|^k \geqslant m^{1-k} \left(\sum_{i=1}^{m} |x_i|\right)^k, \ for\ k < 1 \tag{3.10}$$

Lemma 3.4 *[22] For the RBFNN described as (3.2), the regression function* $S(z)$ *has a certain upper bound for any input as:*

$$\|S(z)\| \leqslant \sum_{j=0}^{\infty} 3k\,(j+2)^{k-1}\, exp\left(\frac{-2p^2 j^2}{\sigma^2}\right) := \bar{S} \tag{3.11}$$

where p *is defined as:* $p = \frac{1}{2} min\,\|\zeta_x - \zeta_y\|$ *where* $x \neq y$ *and* ζ *and* σ *represent the nodes' set and variance while* k *denotes the dimension of* z*. It should be noted that the positive constat* $\bar{S}$ *is independent of input* z*.*

Remark 3.1 In order to express the process more succinctly, a function is designed as follows:

$$SIG(x) = \sqrt[r]{\left[|x_1|^r \cdot sign(x_1), \ldots, |x_n|^r \cdot sign(x_n)\right]^T} \tag{3.12}$$

where $sign(\cdot)$ represents the signum function, $x = [x_1, \ldots, x_n]^T$, $i = 1, \ldots, n$ and $r > 0$ is a constant parameter.

3.2.4 Force Observer Setup

Define the momentum of the robot as:

$$p_m = M(q)\,\dot{q} \tag{3.13}$$

where $p_m \in \mathbb{R}^n$. Then based on (3.13), the derivative of momentum p_m can be written as:

$$\begin{aligned}\dot{p}_m &= \dot{M}(q)\dot{q} - C(q,\dot{q})\dot{q} - G(q) + \tau_C + \tau_{Ex}\\ &= C^T(q,\dot{q})\dot{q} - G(q) + \tau_C + \tau_{Ex}\end{aligned} \tag{3.14}$$

where the i-th component of the n-dimension vector $\dot{p}_m$ is defined as:

$$\dot{p}_{m_i} = -\frac{1}{2}\dot{q}^T\frac{\partial M(q)}{\partial q_i}\dot{q} - G_i(q) + \tau_{C_i} + \tau_{Ex_i} \tag{3.15}$$

where $i = 1, 2, 3, \ldots$. It could be seen that each component of $\dot{p}_{m_1}$ is only affected by the corresponding torque component.

The observer monitoring method based on generalized momentum proposed in [23, 24] is designed to avoid the inversion of the inertia matrix of the robot. Furthermore, it can decouple the estimation results and avoid the estimation of joint acceleration.

Firstly, define a matrix as:

$$\begin{aligned}\Gamma &:= C(q,\dot{q})\dot{q} - \dot{M}(q)\dot{q} + G(q)\\ &= -C^T(q,\dot{q})\dot{q} + G(q)\end{aligned} \tag{3.16}$$

Then the dynamic equations of the momentum observer can be written as:

$$\begin{aligned}\dot{\hat{p}} &= \tau_C - \hat{\Gamma}(q,\dot{q}) + \sigma\\ \dot{\sigma} &= K_{Ex}\left(\dot{p} - \dot{\hat{p}}\right)\end{aligned} \tag{3.17}$$

where matrix $K_{Ex} = diag\left\{k_{Ex_i}\right\} > 0$ and vector $\sigma \in \mathbb{R}^m$ represent the gain matrix and the output of the momentum observer and $\hat{p}$ is the observed momentum. Then the output σ can be further defined as:

$$\begin{aligned}\sigma &= K_{Ex}\left(p(t) - \int_0^t \dot{\hat{p}}(k)dk - p(0)\right)\\ &= K_{Ex}\left(p(t) - \int_0^t \left(\tau_C - \hat{\Gamma}(q,\dot{q}) + \sigma\right)dk - p(0)\right)\end{aligned} \tag{3.18}$$

where $p = \hat{H}(q)\dot{q}$. Taking $\hat{H} = H$ and $\hat{\Gamma} = \Gamma$, the dynamic relationship between the external torque applying on the manipulator and the output of the momentum observer can be written as:

$$\dot{\sigma} = K_{Ex}(\tau_{Ex} - \sigma) \tag{3.19}$$

With the help of Laplace transformation, the i-th component of output σ can be written as:

$$\sigma_i = \frac{K_{Ex_i}}{s + K_{Ex_i}}\tau_{Ex_i} = \frac{1}{1 + T_{Ex_i}s}\tau_{Ex_i},\ i = 1, 2, \ldots, m \tag{3.20}$$

Fig. 3.1 Schematic diagram of force observer

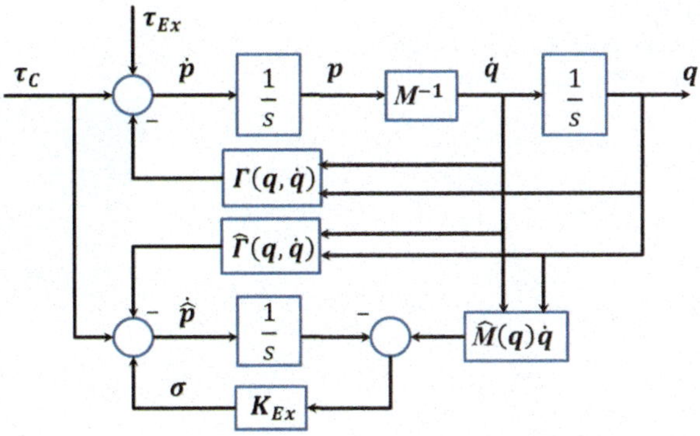

where the time constant $T_{Ex_i} = 1/K_{Ex_i}$ meaning that in the transient response of the i component of output σ, a larger value of K_{Ex_i} would produce a smaller time constant T_{Ex_i}, where output σ_i is related to the same component of the external torque τ_{Ex_i}. According to theoretical analysis, it can be seen that with $K_{Ex} \rightarrow \infty$ it can get $\sigma \approx \tau_{Ex}$. However, an infinite large gain is unreachable while it would cause instability to the system which should be avoided. Therefore, the appropriate value can only be obtained through continuous testing (Fig. 3.1).

Based on the above characteristics, it is possible to accurately estimate $\hat{\tau}_{Ex}$, the external torque applied to the robotic arm, where the external force F_e can be obtained with the help of the Jacobian matrix of the robotic arm.

3.2.5 RCM Constraint

For a manipulator operating in the real world, the task space is a six dimensions Cartesian space, which means $m = 6$. Therefore, the operating coordinate $X = [X_P, X_R]^T \in \mathbb{R}^6$ represents the actual end-effector posture, including the actual end-effector position $X_P = [x, y, z]$ and the actual end-effector orientation $X_R = \left[\alpha, \beta, \gamma\right]$ which is expressed by Euler Angles. Vector $X_D = \left[X_{D_P}, X_{D_R}\right] \in \mathbb{R}^6$ represents the desired end-effector position, where $X_{D_P} = [x_D, y_D, z_D]$ represents the desired end-effector Cartesian position given by the operator, and the desired end-effector orientation $X_{D_R} = \left[\alpha_D, \beta_D, \gamma_D\right]$ is calculated online based on X_P to ensure that the end-effector as well as the manipulator keep close to the RCM position $P_R = [x_R, y_R, z_R]$ during the task.

According to Fig. 3.2, ω is defined as the angle between the actual posture attitude $\overrightarrow{\varrho}$ and the desired posture attitude $\overrightarrow{\varrho_D}$,which could be calculated as $\omega = arctan\frac{\overrightarrow{\varrho_D}\times\overrightarrow{\varrho}}{\overrightarrow{\varrho_D}\cdot\overrightarrow{\varrho}}$. In order to describe the rotation of from $\overrightarrow{\varrho}$ to $\overrightarrow{\varrho_D}$, the unit rotation vector $\overrightarrow{\varrho_U} = \left[\varrho_x, \varrho_y, \varrho_z\right]$ can be calculated through: $\overrightarrow{\varrho_U} = \frac{\overrightarrow{\varrho_D}\times\overrightarrow{\varrho}}{\|\overrightarrow{\varrho_D}\times\overrightarrow{\varrho}\|}$. Then the desired end-effector orientation can be calculated through the following equation:

Fig. 3.2 Schematic of maintaining RCM constraint in the main task space

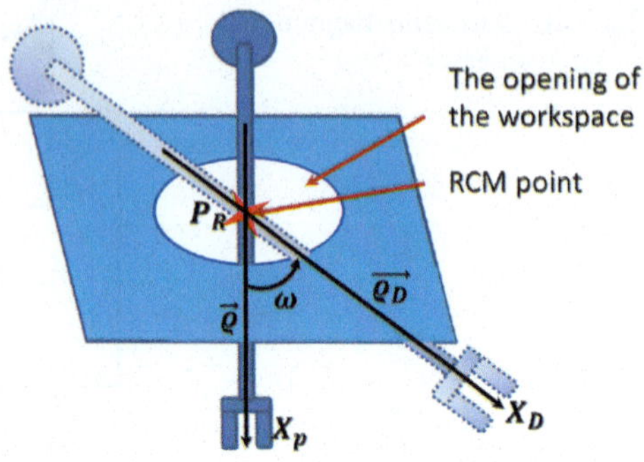

$$\Theta_D = I + \Lambda \cdot \sin(\omega) + 2 \cdot \Lambda^2 \sin^2 \cdot \left(\frac{\omega}{2}\right) \cdot \Theta \tag{3.21}$$

where Θ and Θ_D are the actual and desired rotation matrices respectively, and with the help of Euler transformation X_{D_R} and X_R can be attained. And Λ is the transfer matrix defined as:

$$\Lambda = \begin{bmatrix} 0 & -\varrho_x & \varrho_y \\ \varrho_z & 0 & \varrho_x \\ -\varrho_y & \varrho_x & 0 \end{bmatrix} \tag{3.22}$$

3.3 Controller Design

3.3.1 Fixed-Time Controller

Step 1: In order to reach the desired end-effector posture and achieve the guaranteed performance, a fixed-time controller strategy is inspired. The manipulator system described by (3.1) can be transformed as:

$$\begin{aligned} \dot{x}_1 &= x_2, \\ \dot{x}_2 &= M^{-1}(q)\left(-\left[C(q, \dot{q})x_2 + G(q)\right] + \tau\right) \end{aligned} \tag{3.23}$$

where x_1 and x_2 are defined as $x_1 = [x_{11}, \ldots, x_{xn}]^T \in \mathbb{R}^n$ and $x_2 = [x_{21}, \ldots, x_{2n}]^T \in \mathbb{R}^n$ respectively, where $x_{1i} = q_i, x_{2i} = \dot{q}_i, i = 1, \ldots, n$. Then the error signals could be formulated as:

$$\begin{aligned} e_1 &= x_1 - q_d, \\ e_2 &= x_2 - x_{2d} \end{aligned} \tag{3.24}$$

where x_{2d} is the virtual control quantity designed as:

$$x_{2d} = -K_{11}e_1 - K_{12} \cdot SIG^{\alpha}(e_1) - K_{13} \cdot SIG^{\beta}(e_1) + \dot{q}_d + e_2 \tag{3.25}$$

where the parameters would be designed following: $0 < \alpha < 1$, $\beta > 1$ and $K_{1k} = diag\left\{k_{1k,i}\right\}$ where $k_{1k,i} > 0$ and $j = 1, 2, 3, i = 1, \ldots, n$.

Since the dynamic parameters of the manipulator are unknown, we used the incremental RBFNN method as (3.2) to estimate the parameters as follows:

$$- M \cdot \dot{x}_{2d} - C \cdot x_{2d} - G = W^T S(z) + \epsilon(z) \tag{3.26}$$

where the input z is formulated as $z = \left[q^T, \dot{q}^T, {x_{2d}}^T, \dot{x}_{2d}^T\right]$.

In order to realize the fixed-time convergence control, an adaptive control strategy is designed as follows:

$$\tau = -e_1 - K_{21}e_2 - K_{22} \cdot SIG^{\alpha}(e_2) - K_{23} \cdot SIG^{\beta}(e_2) - \widehat{W}^T S(z) \tag{3.27}$$

where the diagonal weighting matrixes are designed as $K_{2k} = diag\left\{k_{1k,i}\right\}$ with $k_{21,i} > \frac{1}{2}$ and $k_{2k,i} > 0$, $k = 2, 3$, $i = 1, \ldots, n$. It should be claiming that $\widehat{W}$ is the estimation of W while $e_W = W - \widehat{W}$ represents the estimation error.

The updated law of the weighting matrix using the incremental learning neural network is deigned as:

$$\dot{\widehat{W}} = \Gamma \cdot \left(S(z)\, {e_2}^T - \lambda \widehat{W}\right) \tag{3.28}$$

where the diagnoal weighting matrix Γ is defined as:$\Gamma = diag\{\Gamma_i\}$ with $\Gamma_i > 0$ and $\lambda > 0$.

3.3.2 RCM Constraint in Null Space

In traditional manipulator control methods, for solving the problem of limitations, complex mathematical methods are usually used, which require an accurate model of the environment and also need to solve problems such as the non-existence of the inverse matrix. Inspired be the previous work, we propose an RCM constraint method based on the null-space swivel-motion control strategy, which reduces the complexity of calculation improving the timeliness.

It should be noted that different from the RCM method used in minimally invasive surgery, whose end-effector is equipped with special actuators giving it more freedom and more maneuverability, for most industrial scenarios, the last one or two joints and links would have to operate through the open of workspace, which would reduce the redundancy of redundant manipulator, this limitation causes the challenge for industrial RCM control. In this section, we adapt a null space control strategy to accomplish the RCM constraint control task.

The purpose of satisfying the RCM constraint is designed as the control goal of keeping the manipulator as close to the RCM position as possible while avoiding the impact on the main task of grasping the target with the desired end-effector posture. Therefore for a m-joints manipulator operating in the n-dimension space, the null-space sub-mission control torque is designed as follows:

$$\tau_{null} = U \cdot \tau_{sub} = \left[I - J(q)^T \cdot \left(J(q)^{\natural} \right)^T \right] \cdot \tau_{sub} \tag{3.29}$$

where $\tau_{sub} \in \mathbb{R}^n$ represents the subtask control torque acting directly on the manipulator and $U \in \mathbb{R}^{m \times n}$ represents the null-space matrix which could transfer the subtask torque into the decoupled null-space control torque. $J(q)^{\natural} \in \mathbb{R}^{n \times n}$ is the weighted generalized inverse Jacobian matrix, denoted as $J(q)^{\natural} = S^{-1} J^T \left(J S^{-1} J^T \right)^{-1}$, where $S \in \mathbb{R}^{m \times m}$ is a symmetric positive definite matrix. It should be mentioned that the weighted matrix S mentioned above is designed as: $S = H(q)$, so that the transient and steady states of the manipulator would not interfere with each other, guaranteeing the dynamic consistency of the manipulator.

One of the most important reasons for us adopting the null space control method is that the null space control method could effectively achieve the lower priority subtasks simultaneously while ensuring that the primary task is reached.

In our research, our control method would be carried on the Baxter robot which is a redundant robot with $n = 7$ DoFs. For the control strategy, the sub-mission is designed to satisfy the RCM constraint, therefore the sub-mission control torque $\tau_s \in \mathbb{R}^7$ is defined as:

$$\tau_{sub} = \begin{bmatrix} 0 & 0 & 0 & \left(J_R^T F_R \right)^T \end{bmatrix}^T - \hat{\tau}_{ext} \tag{3.30}$$

where $J_R \in \mathbb{R}^{3 \times 3}$ and $F_R \in \mathbb{R}^3$ respectively represent the partial Jacobin matrix of the last three joints as shown in Fig. 3.3 and the RCM constraint force to pull the link of the manipulator as close as possible to the RCM position. The definition of F_R is shown as follows:

$$\begin{aligned} F_R &= K_R \vec{R}, \\ \vec{R} &= \left(\overrightarrow{P_R P_E} - \overrightarrow{P_{EA} P_E} \right) \cdot \cos(\theta_R) \cdot \frac{\| \overrightarrow{P_R P_E} \|}{\| \overrightarrow{P_{EA} P_E} \|} \end{aligned} \tag{3.31}$$

where vector $\vec{R}$ is shown in Fig. 3.3 and to calculate $\cos(\theta_R)$ would need the help of Heron's formula and the forward kinematics of the manipulator. Dis_R represents the shortest distance between the RCM position and the manipulator, which could be calculated through Heron's formula once the position of RCM is obtained.

3.3.3 Stability Analysis

It has been proved that for a n-DoF redundant manipulator system (3.1), the null space control torque would not affect its stability performance. Adopting the control law of (3.27), the Lyapunov function is designed as follows:

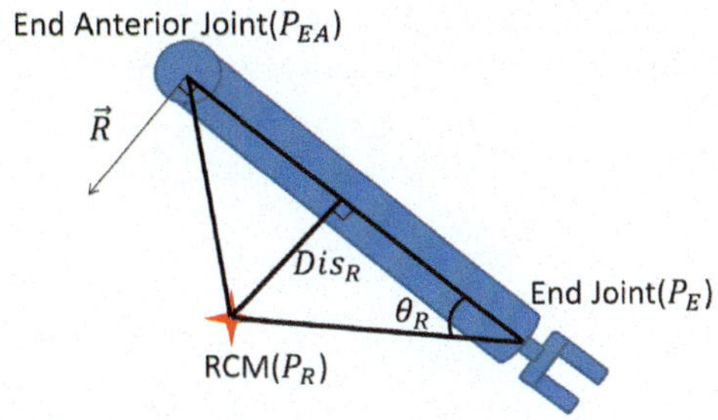

Fig. 3.3 Schematic of RCM, end joint and end anterior joint

$$V = V_1 + V_2 + V_3 \tag{3.32}$$

with:

$$\begin{aligned} V_1 &= \frac{1}{2} {e_1}^T \cdot e_1 \\ V_2 &= \frac{1}{2} {e_2}^T M e_2 \\ V_3 &= \frac{1}{2} tr\left({e_W}^T \Gamma^{-1} e_W\right) \end{aligned} \tag{3.33}$$

Taking the time derivation of V_1 and combine (3.24) and (3.25), we have:

$$\begin{aligned} \dot{V}_1 &= {e_1}^T \cdot \dot{e}_1 \\ &= {e_1}^T \left(e_2 - K_{11}e_1 - K_{12} \cdot SIG^{\alpha}(e_1) - K_{13} \cdot SIG^{\beta}(e_1)\right) \\ &\leqslant {e_1}^T e_2 - \sum_{i=1}^{n} \left(k_{11,i} \cdot e_{1,i}^2\right) - \sum_{i=1}^{n} \left(k_{12,i} \cdot \left|e_{1,i}\right|^{\alpha+1}\right) \\ &- \sum_{i=1}^{n} \left(k_{13,i} \cdot \left|e_{1,i}\right|^{\beta+1}\right) \end{aligned} \tag{3.34}$$

Taking the time derivation of V_1 and combine (3.27), we have:

$$\begin{aligned} \dot{V}_2 &= {e_2}^T M \dot{e}_2 + \frac{1}{2} {e_2}^T \dot{M} e_2 \\ &= {e_2}^T \left(-e_1 - K_{12}e_2 - K_{22} \cdot SIG^{\alpha}(e_2)\right. \\ &\quad \left. -K_{23} \cdot SIG^{\beta}(e_2) + {e_W}^T + \epsilon\right) \\ &= -{e_2}^T e_1 - \sum_{i=1}^{n} \left(k_{21,i} \cdot e_{2,i}^2\right) - \sum_{i=1}^{n} \left(k_{22,i} \cdot \left|_{2,i}\right|^{\alpha+1}\right) \\ &\quad - \sum_{i=1}^{n} \left(k_{23,i} \cdot \left|_{2,i}\right|^{\beta+1}\right) + {2_2}^T {e_W}^T S(z) + {e_2}^T \epsilon \end{aligned} \tag{3.35}$$

Taking the time derivation of V_3 and combining the update law (3.28) with the help of Young's inequality, we have:

$$
\begin{aligned}
\dot{V}_3 &= -tr\left(e_W \Gamma^{-1} \dot{\widehat{W}}\right) \\
&= -tr\left(e_W \Gamma^{-1} \Gamma \cdot \left(S(z)\, {e_2}^T - \lambda \widehat{W}\right)\right) \\
&= -tr\left({e_W}^T S(z)\, {e_2}^T\right) + \lambda \cdot tr\left({e_W}^T \widehat{W}\right) \\
&\leqslant -{e_2}^T e_W S(z) - \frac{\lambda}{2} \cdot tr\left({e_W}^T e_W\right) + \frac{\lambda}{2} \cdot tr\left(W^T W\right)
\end{aligned}
\tag{3.36}
$$

Therefore the time derivation of V could be written as:

$$
\begin{aligned}
\dot{V} &= \dot{V}_1 + \dot{V}_2 + \dot{V}_3 \\
&\leqslant -\sum_{i=1}^{n} \left(k_{11,i} \cdot e_{1,i}^2\right) - \sum_{i=1}^{n} \left(k_{21,i} \cdot e_{2,i}^2\right) + {e_2}^T \epsilon \\
&\quad - \frac{\lambda}{2} \cdot tr\left({e_W}^T e_W\right) + \frac{\lambda}{2} \cdot tr\left(W^T W\right) \\
&\leqslant -\sum_{i=1}^{n} \left(k_{11,i} \cdot e_{1,i}^2\right) - \sum_{i=1}^{n} \left(k_{\bar{21},i} \cdot e_{2,i}^2\right) \\
&\quad + \frac{\epsilon^2}{2} - \frac{\lambda}{2} \cdot tr\left({e_W}^T e_W\right) + \frac{\lambda}{2} \cdot tr\left(W^T W\right)
\end{aligned}
\tag{3.37}
$$

where $k_{\bar{21},i} = k_{21,i} - 0.5$.

Taking $q = \frac{\epsilon^2}{2} + \frac{\lambda}{2} \cdot tr\left(W^T W\right)$, which is obviously positive, and $p = \left(min(2\lambda_{min}\left(K_{11}\right), \frac{2\lambda_{min}\left(K_{\bar{21}}\right)}{\lambda_{max}(H)}, \frac{2\lambda}{\lambda_{max}\left(\Gamma^{-1}\right)}\right)$, where $\lambda_{max}(\cdot)$ and $\lambda_{min}(\cdot)$ represent the maximum and minimum value of the matrix eigenvalues within respectively. Then we have:

$$
\dot{V} < -pV + q \tag{3.38}
$$

then according to Lemma 3.1, the Lyapunov function (3.32) would be UUB as long as the initial state $V(0)$ is bounded.

Taking another Lyapunov function as:

$$
V' = V_1 + V_2 \tag{3.39}
$$

and derivate it over time with the help of Lemma 3.3 and Lemma 3.4, we have:

$$
\begin{aligned}
\dot{V}' &= \dot{V}_1 + \dot{V}_2 \\
&\leqslant -\sum_{i=1}^{n}\left(k_{12,i}\cdot\left|e_{1,i}\right|^{\alpha+1}\right) - \sum_{i=1}^{n}\left(k_{13,i}\left|e_{1,i}\cdot\right|^{\beta+1}\right) \\
&\quad -\sum_{i=1}^{n}\left(k_{22,i}\cdot\left|e_{1,i}\right|^{\alpha+1}\right) - \sum_{i=1}^{n}\left(k_{23,i}\left|e_{1,i}\cdot\right|^{\beta+1}\right) \\
&\quad + {e_2}^T {e_W}^T S(z) + {e_2}^T \epsilon \\
&\leqslant -\lambda_{min}(K_{12})\sum_{i=1}^{n}\left(\left|e_{1,i}\right|^{\alpha+1}\right) - \lambda_{min}(K_{13})\sum_{i=1}^{n}\left(\left|e_{1,i}\right|^{\beta+1}\right) \\
&\quad - \lambda_{min}(K_{22})\sum_{i=1}^{n}\left(\left|e_{2,i}\right|^{\alpha+1}\right) - \lambda_{min}(K_{23})\sum_{i=1}^{n}\left(\left|e_{2,i}\right|^{\beta+1}\right) \\
&\quad + {e_2}^T {e_W}^T S(z) + {e_2}^T \epsilon
\end{aligned}
\tag{3.40}
$$

Since according to Lemma 3.4, e_2, e_W, $S(z)$ and ϵ are bounded, denoting $\bar{U}_1$ and $\bar{U}_2$ as the upper bound of $\left\|{e_2}^T {e_W}^T S(z)\right\|$ and $\left\|e_2^T \epsilon\right\|$ respectively. Design κ_1 and κ_2 as follows:

$$\kappa_1 = min\left(2\lambda_{min}(K_{12}), 2\lambda_{min}(K_{22})\right) \tag{3.41}$$

$$\kappa_2 = min\left(\frac{2\lambda_{min}(K_{12})\, n^{\frac{1-\beta}{2}}}{\lambda_{max}(H)}, \frac{2\lambda_{min}(K_{23})\, n^{\frac{1-\beta}{2}}}{\lambda_{max}(H)}\right) \tag{3.42}$$

then we have:

$$\dot{V}' \leqslant -\kappa_1 V'^{\frac{\alpha+1}{2}} - \kappa_2 V'^{\frac{\beta+1}{2}} + \nu \tag{3.43}$$

where $\nu = \bar{U}_1 + \bar{U}_2$. According to Lemma 3.2, the correlation signal $[e_1, e_2]$ would converge to the compact set Ω in fixed time as $0 \leqslant T \leqslant T_{max} = \frac{2}{\kappa_1\gamma(1-\alpha)} + \frac{2}{\kappa_2\gamma(\beta-1)}$.

3.4 Grasping Detection and RCM Position Recognition

In this section, we would introduce the grasp position and orientation detection method used for getting the desired end-effector position for the redundant manipulator executing the grasping task.

3.4.1 Problem Description of Grasp Detection

The input of our method is the RGB image captured from the desktop scene and the aligned depth image, and the output is based on the five-dimensional grasping rectangle proposed in [25], which can be converted into grasping posture through hand-to-eye or hand-to-eye calibration, and the specific form is shown as follows:

$$g = \{x, y, \Phi, h, w\} \tag{3.44}$$

where g represents the grabbing rectangle, defined by the end-gripper, (x, y) and Φ represents the center position and orientation of the grasping rectangle respectively, and with these three parameters, the grasping posture for the target of the end-gripper can be calculated. Parameters h and w respectively represent the thickness and open width of the end-gripper.

Since in our experiment, the manipulator we use is the right arm of the Baxter robot, whose end-gripper has a fixed value for its thickness, which means that we can ignore the parameter h. Also, in order to use the rectangle representation, the numerical relationship between h and w is designed as: $w = 4 \times h$.

The improved morphological image processing algorithm proposed in previous work [26] is used to choose the grasping rectangle g. For each g, the feature set F is generated as $F = \{f_1, f_2, f_3, \ldots, f_r\}$, which consists of different information including RGB color information, depth information of each point, surface normal information, and so on. Based on these features, a CNN classifier is adopted to score and classify the whole feature set, and determine the best grasping configuration.

3.4.2 CNN Classification for Grasping Posture

3.4.2.1 Image Rre-processing

The classification model is trained and evaluated based on the Cornell Grasping Dataset. Inspired by [7], then the CNN classifier is designed with 7 channels of inputs, and the image information is extracted from a 24×24 rectangle, which includes RGB space, depth and surface normal data. Since the RGB data and surface normal data are three dimension data, the dimension of input is rescaled to $24 \times 24 \times 7$ with a rescale operation. Also, in order to obtain better performance on the CNN network, every channel is normalized by subtracting its mean value and deviating from its standard deviation before the network was introduced, due to the difference between RGB information and other information including depth information and surface normal information.

3.4.2.2 Grasping Classification

In previous work, the CNN classifier only considered the end-effector with a straight down posture and without considering the boundaries of the workspace, this might not work well for some cases. For example, the grasp target close to the boundary of the workspace, which would result in the collision with the workspace boundary during grasping action with the straight down end posture. In this case, the end-effector posture needs to tilt to grasp the target without collision. Therefore, a safety region method is adopted in our new CNN classifier as follows:

$$gt = \begin{cases} \{1, 0, 0\}, & positive\ grasp\ action\ within\ safety\ region \\ \{0, 1, 0\}, & positive\ grasp\ action\ outside\ safety\ region \\ \{0, 0, 1\}, & negative\ grasp\ action. \end{cases} \tag{3.45}$$

3.4.2.3 Training Progress

In this part, the loss function is designed with the cross entropy method, which could improve the performance of the network, and the $24 \times 24 \times 7$ feature is rescaled to $28 \times 28 \times 7$ matching the input of the safety region CNN network designed above. A single NVIDIA RTX2080 graphics with 8G-Bits of GPU memory is used to train the CNN network described above end-to-end. Then based on the PyTorch with cuda-10.0 and cudnn-7.4.2. the framework is tested. During the training process, the learning rate is set as 0.001 with the Adam optimizer and the number of training epochs is set as 25, then the accuracy of the training phase reached 98%.

3.5 Experiments

In this Section, the feasibility of the proposed RCM control strategy in the previous section is proved through a few grasping experiments based on a Baxter robot. A schematic of the workspace is shown in the Fig. 3.4. Considering the volume of the Baxter manipulator, the workspace is designed as a rectangular space of $0.5 \times 0.37 \times 0.3\ \text{m}^3$. Using the similar background remove method of the grasp detection problem, the edge of the opening of the workspace can be recognized. Then the position of the RCM point, defined as the center position of the opening of the workspace, can be obtained through a few mathematical calculations. A video containing some representative experiments is attached and uploaded.

3.5.1 Experimental System Setup

To verify the proposed redundant manipulator RCM constraint control strategy, a $n = 7$ redundant Baxter robot is adapted, and the visual grasping detection part is carried out based on a Kinect camera.

The experimental system uses two computers, in which the upper computer is responsible for receiving the data of the Baxter robot and the data of the lower computer, estimating the external force based on the force observer, running the

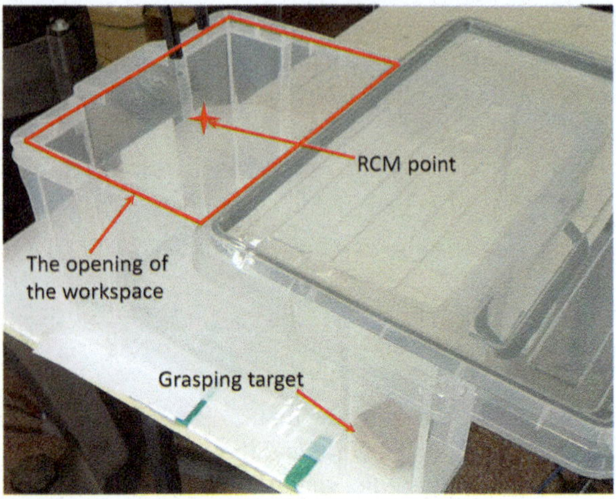

Fig. 3.4 Workspace with a small opening

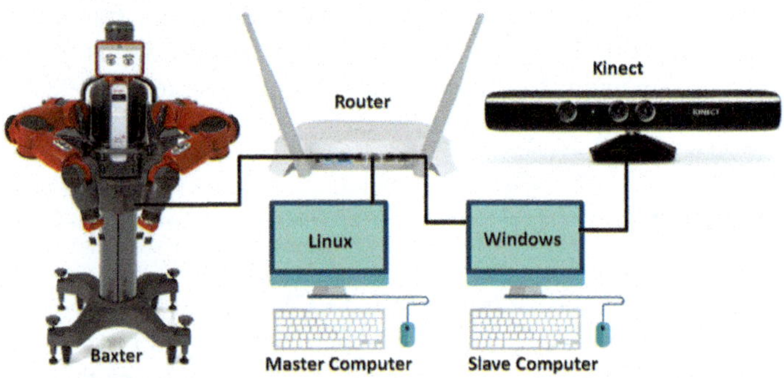

Fig. 3.5 Experimental system composed of Baxter robot, Kinect camera, Linux system computer and Windows system computer

main program and sending control commands to the robot. The lower computer receives the data from the Kinect camera, calculates the RCM position, the position of the expected grabbing point and the grabbing posture, and transmits the calculation results to the upper computer through the User Datagram Protocol (UDP). The whole experimental system is shown in Fig. 3.5.

3.5.2 Grasping Task with Fixed RCM Point

In the experiment, the task of the Baxter robot successfully grasping the target cube in a small opening workspace without collision is designed as the four steps as shown in Fig. 3.6. Firstly, Fig. 3.6a shows that in order to make the experiment universal

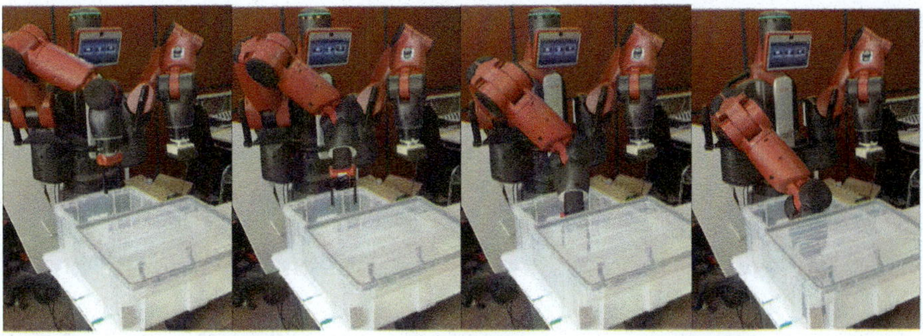

(a) First step of experiment. (b) Second step of experiment. (c) Third step of experiment. (d) Fourth step of experiment.

Fig. 3.6 Steps diagram of Baxter performing grasping task

and repeatable, the initial posture of the Baxter robot is designed as the untuck pose. Secondly, as shown in Fig. 3.6b, the right arm of the Baxter robot would move the end-effector to reach the RCM position with the desired posture based on the RCM position and the expected grasping position obtained from the Kinect camera connected to the lower computer. Then Fig. 3.6c shows the process of Baxter moving the end-effector towards the desired grasping position subjecting to the RCM constraint. Finally, as shown in Fig. 3.6d, Baxter robot reaches the desired grasping position and grasps the cube successfully. Through the experiments, the weighting matrixes for control are selected as: $K_{11} = diag\{30\}$, $K_{12} = diag\{2\}$, $K_{13} = diag\{2\}$, $K_{21} = diag\{20\}$, $K_{22} = diag\left\{\frac{3}{2}\right\}$, $K_{23} = diag\left\{\frac{3}{2}\right\}$; the RBFNN parameters are selected as: $\varepsilon = 10, \delta = 0.9, m = 3, \sigma = 7.1$; the fixed-time parameters are selected as: $\alpha = \frac{1}{2}, \beta = 2$.

To verify the feasibility and applicability of the RCM control strategy proposed in this article, four grasping experiments with fixed RCM positions are conducted based on the right arm of the Baxter robot, the results are shown in Figs. 3.7, 3.8, 3.9 and 3.10 respectively, where $\widetilde{X}(q)$ represents the distance between the end-effector of Baxter robot's right arm and the grasping target and Dis_R represents the shortest distance between the RCM position and Baxter robot's right arm. In these experiments, the grasping target is placed in different positions inside the workspace with different RCM positions. Through the visual localization algorithm, the three-dimensional coordinates of RCM position and desired grasping position are obtained respectively as follows: For the first experiment shown in Fig. 3.7: $P_R = [0.59, -0.27, 0.10]$, $X_D = [0.72, 0.00, -0.16]$; for the second experiment shown in Fig. 3.8: $P_R = [0.70, -0.25, 0.10]$, $X_D = [0.96, -0.15, -0.16]$; for the third experiment shown in Fig. 3.9: $P_R = [0.70, -0.22, 0.10]$, $X_D = [0.95, -0.35, -0.17]$; for the fourth experiment shown in Fig. 3.10: $P_R = [0.60, -0.35, 0.13]$, $X_D = [0.40, -0.28, -0.15]$.

According to Figs. 3.7, 3.8, 3.9 and 3.10, it can be seen that the RCM constraint null-space control strategy proposed in this section has good control performance,

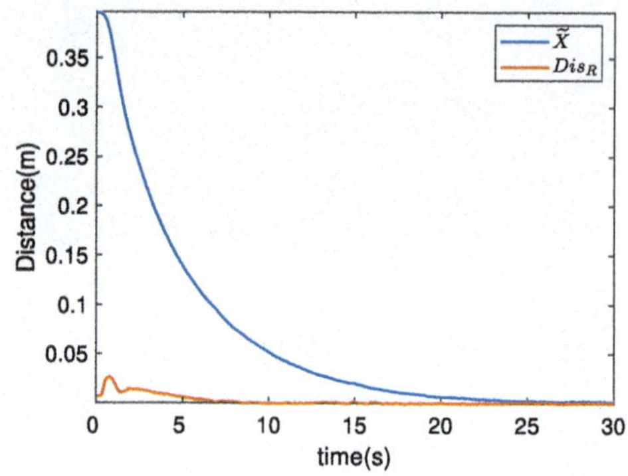

Fig. 3.7 $\widetilde{X}(q)$ and Dis_R in the first experiment

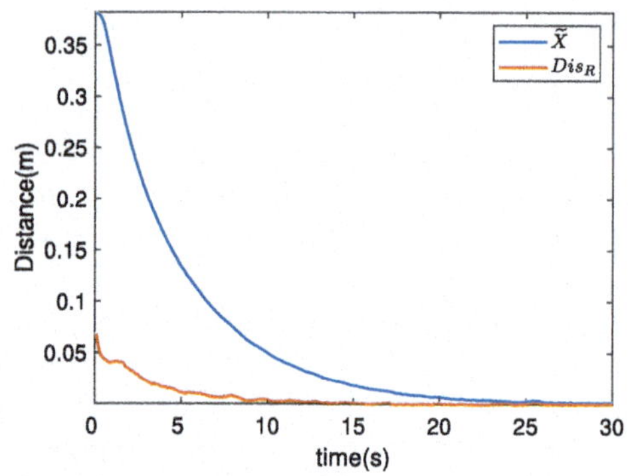

Fig. 3.8 $\widetilde{X}(q)$ and Dis_R in the second experiment

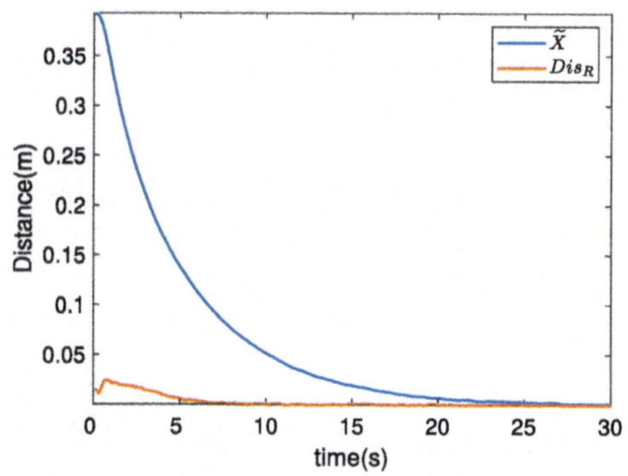

Fig. 3.9 $\widetilde{X}(q)$ and Dis_R in the third experiment

which can not only accomplish the simple task where the RCM position is closer to the Baxter robot arm base than the grasping target position in the first three experiments, but also accomplish the more difficult task in the fourth experiment where the RCM position is further away from the Baxter robot arm base than the grasping target.

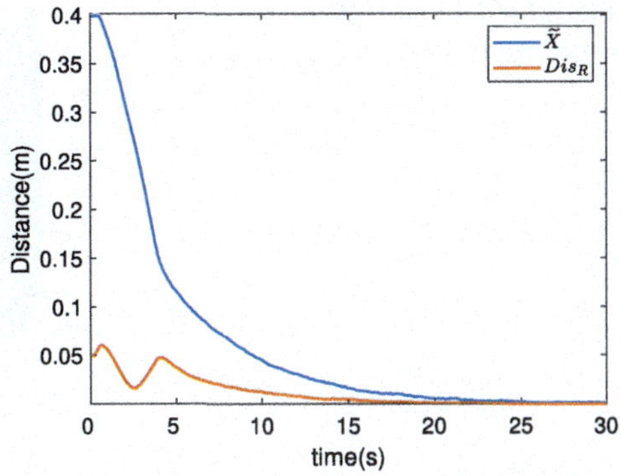

Fig. 3.10 $\widetilde{X}(q)$ and Dis_R in the forth experiment

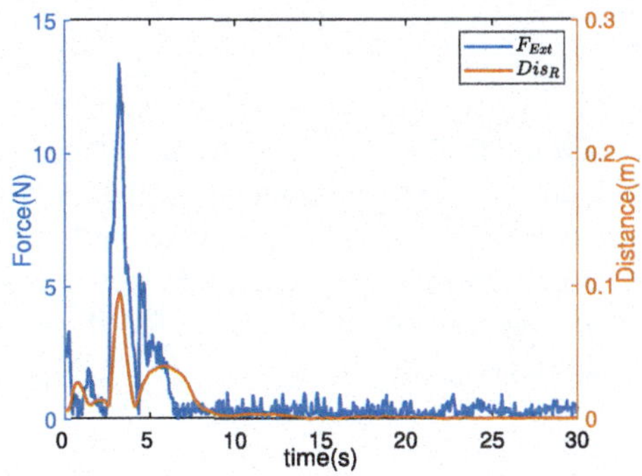

Fig. 3.11 The observed external force and Dis_R

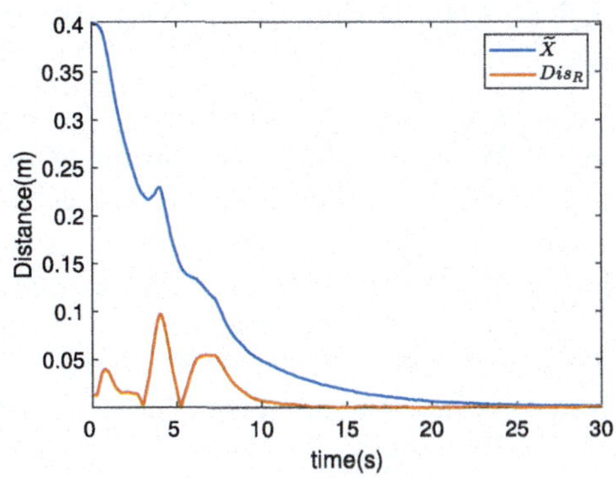

Fig. 3.12 $\widetilde{X}(q)$ and Dis_R in the force contact experiment

3.5.3 Grasp Task with RCM Constrain and Force Contact

After verifying the RCM constraint control strategy, external force is applied to the Baxter robotic with the same RCM position and grasping target position in the first experiment aiming to verify the feasibility of the force observer method and the stability of the null-space control strategy proposed in this section, and the experimental results are shown in Figs. 3.11 and 3.12.

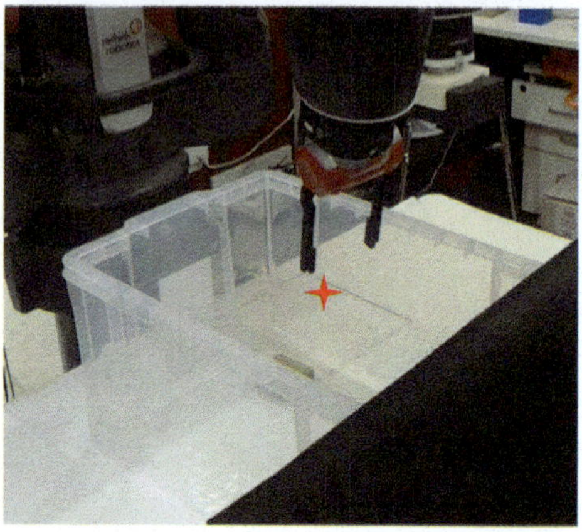

Fig. 3.13 Dynamic RCM experiment at $t = 0$ s

From Fig. 3.11, it can be found that the force observer could quickly and accurately calculate the external force applied to the robot arm. Also, the RCM constraint null-space control algorithm proposed in this section enables the arm to quickly return to an attitude close to the RCM position after a large external force is applied to make the arm deviate from the RCM position, and also reduces the impact on the end-effector position.

3.5.4 Grasp Task with Dynamic RCM Point

Considering that in actual production assembly tasks, in many cases the position of the RCM changes due to the change of the task space opening. In order to verify the capability of the proposed RCM constraint control strategy, the RCM position is changed dynamically by masking the opening of the workspace artificially as shown in Figs. 3.13 and 3.14, where the red star represents the RCM point and the red line represents the trajectory of the dynamic RCM. The complete trajectory of the dynamic RCM is shown in Fig. 3.15 and the experimental result is shown in Fig. 3.16. Based on the variation of the shortest distance between the dynamic RCM position and the Baxter robot's right arm, we can conclude that the control strategy proposed in this section could drive the robot arm accurately and quickly track the dynamic RCM while avoiding collision with the openings in the workspace.

3.6 Conclusion

This chapter developed a novel fixed-time RCM control strategy for redundant manipulators operating in the workspace with a small opening adapting a proposed safety region grasping strategy. The experiment results show that with the proposed control

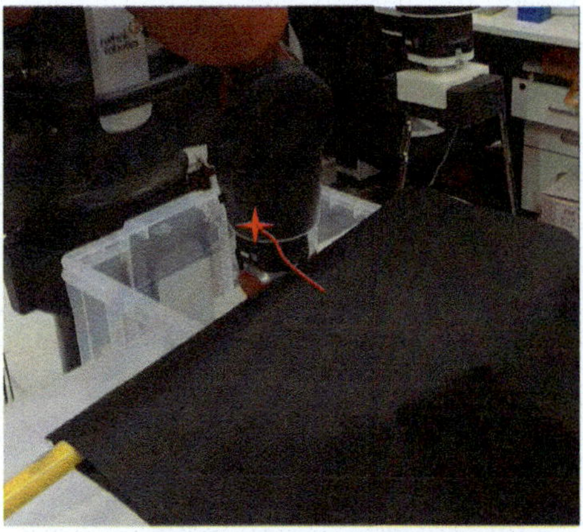

Fig. 3.14 Dynamic RCM experiment at $t = 10$ s

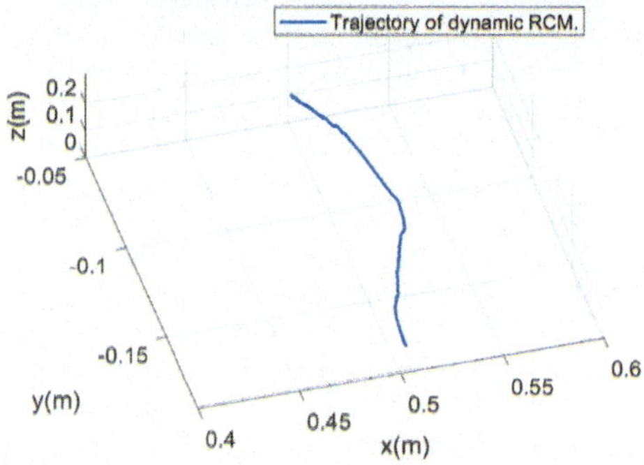

Fig. 3.15 The trajectory of the dynamic RCM

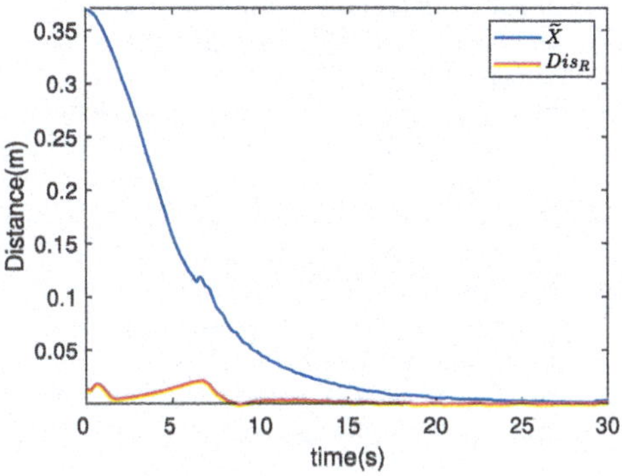

Fig. 3.16 $\widetilde{X}(q)$ and Dis_R in the dynamic RCM experiment

strategy, the redundant manipulator can successfully grasp the target inside the small opening workspace without collision. Also, with the help of a force observer and fast grasping detection algorithm, the proposed control strategy can not only interact smoothly with the surrounding environment as well as the cooperator but also track the dynamic RCM position in real time.

References

1. Zhang Y et al (2017) Adaptive projection neural network for kinematic control of redundant manipulators with unknown physical parameters. IEEE Trans Indus Electron 65(6):4909–4920
2. Tan N et al (2020) Neural-network-based control of wheeled mobile manipulators with unknown kinematic models. In: Paper presented at the 2020 international symposium on autonomous systems (ISAS), Guangzhou, China, 06–08 Dec 2020
3. Klecker S et al (2017) Neuro-inspired reward-based tracking control for robotic manipulators with unknown dynamics. In: Paper presented at the 2017 2nd international conference on robotics and automation engineering, Shanghai, China, 29–31 Dec 2017
4. Yang R et al (2016) RBFNN based adaptive control of uncertain robot manipulators in discrete time. In: Paper presented at the 2016 UKACC 11th international conference on control (CONTROL), Belfast, UK, 31 Aug 2016–02 Sept 2016
5. Yang C et al (2016) Adaptive RBFNN control of robot manipulators with finite-time convergence. In: Paper presented at the 2016 IECON 2016—42nd annual conference of the IEEE Industrial Electronics Society, Florence, Italy, 23–26 Oct 2016
6. Chen CLP, Liu Z (2017) Broad learning system: an effective and efficient incremental learning system without the need for deep architecture. IEEE Trans Neural Netw Learn Syst 29(1):10–24
7. Redmon J, Angelova A (2015) Real-time grasp detection using convolutional neural networks. In: Paper presented at the 2015 IEEE international conference on robotics and automation (ICRA), Seattle, WA, USA, 26–30 May 2015
8. Kumra S, Kanan C (2015) Robotic grasp detection using deep convolutional neural networks. In: Paper presented at the 2017 IEEE/RSJ international conference on intelligent robots and systems (IROS), Vancouver, BC, Canada, 24–28 Sept 2017
9. Wang Z et al (2016) Robot grasp detection using multimodal deep convolutional neural networks. Adv Mechan Eng 8(9):1687814016668077
10. Guo D et al (2017) A hybrid deep architecture for robotic grasp detection. In: Paper presented at the 2017 IEEE international conference on robotics and automation (ICRA), Singapore, 29 May 2017–03 June 2017
11. Zhang Y et al (2009) More illustrative investigation on window-shaped obstacle avoidance of robot manipulators using a simplified LVI-based primal-dual neural network. In: Paper presented at the 2009 international conference on mechatronics and automation, Changchun, China, 09–12 Aug 2009
12. Yu X et al (2016) Obstacle space modeling and moving-window RRT for manipulator motion planning. In: Paper presented at the 2016 IEEE international conference on information and automation (ICIA), Ningbo, China, 01–03 Aug 2016
13. Park H et al (2013) Intuitive peg-in-hole assembly strategy with a compliant manipulator. In: Paper presented at the IEEE ISR 2013, Seoul, Korea (South), 24–26 Oct 2013
14. Nigro M et al (2020) Peg-in-hole using 3D workpiece reconstruction and CNN-based hole detection. In: Paper presented at the 2020 IEEE/RSJ international conference on intelligent robots and systems (IROS), Las Vegas, NV, USA, 24 Oct 2020–24 Jan 2021
15. Su H et al (2018) Safety-enhanced collaborative framework for tele-operated minimally invasive surgery using a 7-DoF torque-controlled robot. Int J Control Autom Syst 16:2915–2923
16. Su H et al (2019) Improved human-robot collaborative control of redundant robot for teleoperated minimally invasive surgery. IEEE Robot Autom Lett 4(2):1447–1453
17. Slotine JE, WeipingLi L (1988) Adaptive manipulator control: a case study. IEEE Trans Automc Control 33(11):995–1003
18. Zuo Z et al (2018) An overview of recent advances in fixed-time cooperative control of multiagent systems. IEEE Trans Indus Inform 14(6):2322–2334
19. Huang D et al (2019) Composite learning enhanced neural control for robot manipulator with output error constraints. IEEE Trans Indus Inform 17(1):209–218
20. Chen Q et al (2020) Neural-network-based adaptive singularity-free fixed-time attitude tracking control for spacecrafts. IEEE Trans Cybern 51(10):5032–5045

21. He C et al (2020) Fixed-time adaptive neural tracking control for a class of uncertain nonlinear pure-feedback systems. IEEE Access 8:28867–28879
22. Wang C et al (2002) An ISS-modular approach for neural control of pure-feedback systems. In: Paper presented at the 2002 international conference on control and automation, 2002. ICCA. Final Program and Book of Abstracts, Xiamen, China, 19–19 Jun 2002
23. De Luca A, Mattone R (2005) Sensorless robot collision detection and hybrid force/motion control. In: Paper presented at the proceedings of the 2005 IEEE international conference on robotics and automation, Barcelona, Spain, 18–22 Apr 2005
24. De Luca A et al (2006) Collision detection and safe reaction with the DLR-III lightweight manipulator arm. In: Paper presented at the 2006 IEEE/RSJ international conference on intelligent robots and systems, Beijing, China, 09–15 Oct 2006
25. Lenz I et al (2015) Deep learning for detecting robotic grasps. Int J Robot Res 34(4–5):705–724
26. Zhang J et al (2020) Robotic grasp detection using effective graspable feature selection and precise classification. In: Paper presented at the 2020 international joint conference on neural networks (IJCNN), Glasgow, UK, 19–24 Jul 2020

Chapter 4
Motor Learning and Generalization Using Broad Learning Adaptive Neural Control

The human neural motor system has the intelligence to learn new skills, and then generalize these skills naturally. However, it is not easy for a robot to demonstrate such intelligent behaviors. Inspired by the neural motor behaviors, a framework of broad learning-based novel adaptive neural control is proposed in this Chapter, such that in the presence of dynamic disturbance, robots can learn a set of basic skills and then generalize these skills to the neighboring movements naturally as our human motor system. This is achieved by incorporating deterministic learning with the broad learning system which can accumulate and reuse the learned knowledge. The broad learning-enabled adaptive neural control has been rigorously established in theory and tested in both simulation and experimental studies. Simulation results and performance of the Baxter robot in the experiments have shown the effectiveness and superiority of the proposed method in comparison to the conventional adaptive neural control.

4.1 Introduction

In recent years, human-like motor training techniques have been introduced into robotics [1–3], aiming to enable robots to think and act in a human-like manner no matter in biomechanics or intelligence aspect. In our daily life, humans have the distinctive ability to accomplish various complex motor tasks by means of being inspired by the fundamental techniques in the past. Notably, it would be ideal to the engineer robot systems to learn and perform in a similar way as our humans.

In the past decades, motor learning in humans has been studied mostly through adaptation to motor tasks. Force and impedance control have been widely implemented to provide a desired adaptive motor control behavior while the robot interacts with the environment [2, 3]. Kadiallah [4] proposed a structure to adapt the multiple movements in the state space based on nonlinear adaptive control using Radial Basis

C. Yang et al., *Advanced Teleoperation and Robot Learning for Dexterous Manipulation*, Springer Tracts in Advanced Robotics 160,
https://doi.org/10.1007/978-3-031-78501-6_4

Function Neural Network (RBFNN) [5]. For the adaptive control system, deterministic learning theory [6–8] is one of the most efficient methods to learn the dynamic system under the satisfaction of persistent excitation (PE) condition [9]. In a local region, it can approximate the dynamic models accurately along the recurrent-like orbit. Deterministic learning can make the neural weights converge to their optimal values and the system will have exponential stability ultimately via localized RBF neural networks [10, 11].

Recently, a novel framework of broad learning system (BLS) based on incremental learning principle has been developed and successfully applied for pattern recognition and off-line classification [12, 13]. Although both BLS and incremental learning are widely studied by researchers, there is barely adaptive neural control research concerning the incremental learning or BLS to our knowledge, which inspires the research of this section. Hence, this work makes a step forward to extend broad learning to cope with task generalization in motor learning problems by integrating with adaptive neural control.

In this context, an adaptive neural control based on a broad learning framework for motor generalization is proposed with the satisfaction of global stability [14, 15] and prescribed tracking performance [16] under the dynamic external disturbance environment. The main contributions is presented as follows.

1. A new class of input vectors can be accommodated through the proposed method by incorporating broad learning to cope with the problems of determining the appropriate number and position of neural nodes for the NN controller.
2. A novel incremental scheme of neural nodes is designed by considering the input vector and information of past nodes in the network, rather than regarding the current input vector as the new neural nodes directly which seems not to be a reasonable choice [17–19].
3. An adaptive switching scheme is developed through a dynamic radius parameter during the backstepping design in order to ensure closed-loop global stability and make the input signs confined within the dynamic compact set in the process of broad learning.
4. The proposed algorithm, which blends deterministic learning [6–8] is able to improve and then reuse the learned knowledge to enable the controller to act in a human-like manner.

4.2 Problem Fundamentals

4.2.1 Broad Learning System

Recently, Broad Learning System, which has evolved from random vector functional-link neural networks [20], has demonstrated its efficient approximation and generation ability in recognition and classification fields. In this framework, the input vectors of the network are described as X and the output variables are denoted as

$Y \in \mathbb{R}^{N\times C}$, where N is the number of feature mapping and C is the dimension of the network's output. Following the input vectors, the n feature mappings have been used to extract the feature of input vectors

$$F_i = \zeta_1(XW_{fi} + \vartheta_{fi}) \quad i = 1, \cdots, n \tag{4.1}$$

where ζ_1 is the transfer function, W_{fi} and ϑ_{fi} are the weight and width which are randomly generated at the initial stage of the network and then remain constant during training. Then, the m enhancement nodes are obtained by the feature mappings

$$H_j = \zeta_2(F^n W_{ej} + \vartheta_{ej}) \quad j = 1, \cdots, m \tag{4.2}$$

where W_{ei} and ϑ_{ei} are the orthonormal basis vectors, ζ_2 is also the transfer function. Eventually, the output of BLS is represented as

$$\begin{aligned} Y &= [F_1, \cdots, F_n|\zeta_2(F^n W_{e1} + \vartheta_{e1}), \cdots, \zeta_2(F^n W_{em} + \vartheta_{em})]W^{m+n} \\ &= [F_1, \cdots, F_n|H_1, \cdots, H_m]W^{m+n} \\ &= [F^n|H^m]W^{m+n} \\ &= G^{m+n}W^{m+n} \end{aligned} \tag{4.3}$$

For the application of classification and recognition, the testing results are usually given beforehand and the neural weight W^{m+n} in BLS can be obtained by solving the pseudo-inverse of $[F^n|H^m]$, i.e. $W^{m+n} = [F^n|H^m]^{\dagger}Y$.

4.2.2 RBFNN with Broad Learning Framework

Previous researches [6, 21] have indicated that RBFNN can approximate any continuous function $f(\Xi)$ with an optimal neural weight W^* as long as the number of node N is large enough, such that

$$f(\Xi) = W^{*T}S(\Xi) + \epsilon(\Xi), \forall \Xi \in \Omega_\Xi \tag{4.4}$$

where Ξ is the input vector of neural networks, $\Omega_\Xi \subset \mathbb{R}^q$ is a compact set and $|\epsilon(\Xi)| < \epsilon^*$ represents the estimated error for each $\epsilon^* > 0$. Here, $S(\Xi) = [s_1(\Xi), \cdots, s_n(\Xi)]$ and the Gaussian function is used as the transform function for the input data as follows:

$$s_i(\Xi) = exp[\frac{-(\Xi - \varphi_i)^T(\Xi - \varphi_i)}{\eta_i^2}] \quad i = 1, \cdots, n \tag{4.5}$$

where φ and η denote the centers and widths of the RBFNN.

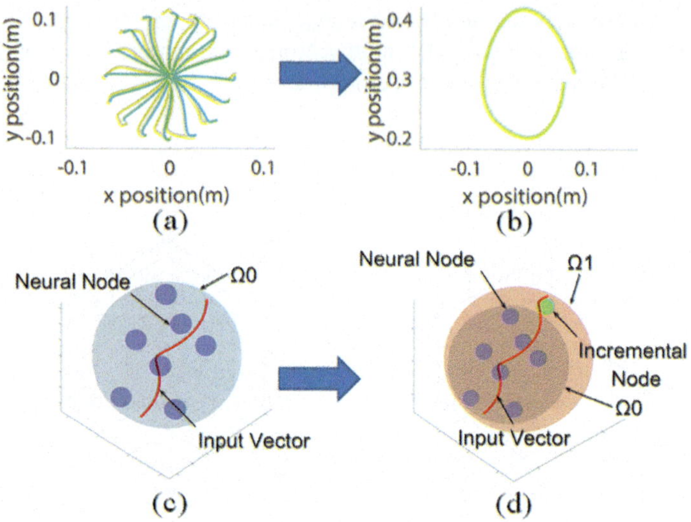

Fig. 4.1 Top panel: motor behaviors in skills learning and transfer [22]. **a** is 8 reaching movements that humans need to adapt to. **b** is cycle-drawing movement which is generalized by the training knowledge from **a**. Bottom panel: RBFNN learning principle, from conventional learning **c** to broad learning with incremental nodes **d**

In this section, a detailed description of the development of RBFNN using a broad learning framework will be given. The inspiration for the proposed framework is presented in Fig. 4.1. As shown in the top panel of Figs. 4.1, the research in [22] reveals that the internal model of the central nervous system (CNS) has the capability of generalization for skills transfer. Inspired by the incremental properties of BLS, the novel RBFNN with broad learning is designed to model motor skills learning and generalization as shown in the bottom panel. The neural network is updated dynamically through the incremental node to accommodate the new input vector, i.e., the position or velocity of the new movement. Specifically, as shown in Fig. 4.1c, while the compact set Ω_0 covers the whole input vector, the neural nodes can be well activated and it may result in good approximation performance. Nevertheless, considering the Gaussian function in the basic function, it can be easily concluded that the input vector may bring an extremely low impact on it if the input vector goes beyond the initial compact set Ω_0 as shown in Fig. 4.1d. In such case, a new neuron will be augmented to the network in order to incorporate the new information and the original compact set Ω_0 will be extended to Ω_1.

Human motor learning has been regarded as an outgrowth of the motor control process. A more recent finding, which inspired this work, is that the novel motor skills learning is associated with the previous knowledge and current environment information [23–25]. At this point, a novel generalized scheme allowing for incrementing the neurons of NN control is proposed in this part. Specifically, φ_{t+T} is calculated by taking into account other existing neural nodes, where T is the calculation interval, rather than using the position of the input data as the neural center directly [17–19]. Assume that the n centers vectors in the existing network centers which are closest to the input vector $\Xi(t)$ are $c_{\min} = \{c_1, c_2, \cdots, c_n\} \in \varphi$, then the parameters of an incremental node can be represented as

$$\varphi_{new} = \bar{c}_{\min} + \beta(\Xi(t) - \bar{c}_{\min}) \tag{4.6}$$

where $\beta \in \mathbb{R}^q$ is a tunable parameter that can determine the position of a new node by considering the input vector and nearest nodes; $\bar{c}_{\min}$, which represents the average position to the closest centers $c_{\min}$, is defined as $\bar{c}_{\min} = \frac{\sum_{i=1}^{n} c_{\min_i}}{n}$. Hence, the new neural center vector can be defined as

$$\varphi(t+T) = \begin{cases} [\varphi(t)\ \varphi_{new}], & if\ dis(\Xi(t), \bar{c}) > \Theta \\ \varphi(t), & otherwise \end{cases} \tag{4.7}$$

where $dis(\Xi(t), \bar{c})$ denotes the norm distance between input vector $\Xi(t)$ and $\bar{c}$, Θ is the predetermined threshold.

Inspired by broad learning system and considering the real-time requirement in the control aspect, a trigonometric expansion scheme [26] is employed to generate the corresponding enhancement nodes. Let k be the current number of RBFNN centers, the enhancement vectors can be represented as

$$H(t) = [H_1, \cdots, H_i] \quad i = 1, \cdots, k \tag{4.8}$$

where $H_i = [\cos(s_i(\Xi(t)), \sin(s_i(\Xi(t))]$. Hence, the hidden layer of the neural network is represented as $G(t) = [S(\Xi)|H]$.

The incremental basic function and corresponding enhancement vector are written as

$$Sn(t+T) = [S(\Xi)|S_{ex}(\Xi)] \tag{4.9}$$

$$Hn(t+T) = [H|\cos(S_{ex}(\Xi)), \sin(S_{ex}(\Xi))] \tag{4.10}$$

$$Wn(t+T) = [W|W_{ex}] \tag{4.11}$$

where $S_{ex}(\Xi)$ is the Gaussian function with the new center φ_{new} of the incremental node. Consequently, the broad RBFNN is defined as

$$\begin{aligned} Y(\Xi) &= [Sn(t+T)|Hn(t+T)]W^T \\ &= Gn * W^T \end{aligned} \tag{4.12}$$

Remark 4.1 Although the research of NN incremental learning has been noticeably increased in the past years [17–19], it is noted that the proposed method also has its unique innovations: (1) In order to fully consider the learned primitive motor knowledge, the parameters of the new neuron are calculated related to the existing nodes of the NN; (2) the number of neural nodes is determined by the distance threshold Θ which user can specify; (3) enhancement nodes are employed to improve the accuracy of approximation and trigonometric expansion scheme is designed to balance the calculation burden; (4) due to the characteristic of the Gaussian function, existing neuron which is far from the new orbit will be less effected during the network training which tend to prevent the learned knowledge.

4.3 Global Adaptive Neural Network Controller Design

Another focus of this work is to equip robots with the capability of skills learning and generalization as human motor systems. In this section, the design of the neural adaptive controller integrated with generalized properties for motor skills learning is elaborated by means of the developed broad RBFNN which is discussed above. Figure 4.2 shows the control diagram of the proposed method.

4.3.1 Prescribed Tracking Performance

The dynamic equations of an n-joint rigid manipulator based on *Lagrange-Euler* form which can be defigsted by [27] is described as

$$M(q)\ddot{x} + C(q, \dot{q})\dot{x} + G(q) = \tau \tag{4.13}$$

where vector $x = [x_1, \ldots, x_n]^T \in \mathbb{R}^n$ and vector $G(q) \in \mathbb{R}^n$ represent n dimensional pose in Cartesian space of the redundant manipulator and gravitational moment respectively, while matrix $M(q) \in \mathbb{R}^{n \times n}$ and matrix $C(q, \dot{q}) \in \mathbb{R}^{n \times n}$ represent inertia matrix and the Coriolis and Centrifugal effects. Vectors $\tau \in \mathbb{R}^n$ denote the control torque input.

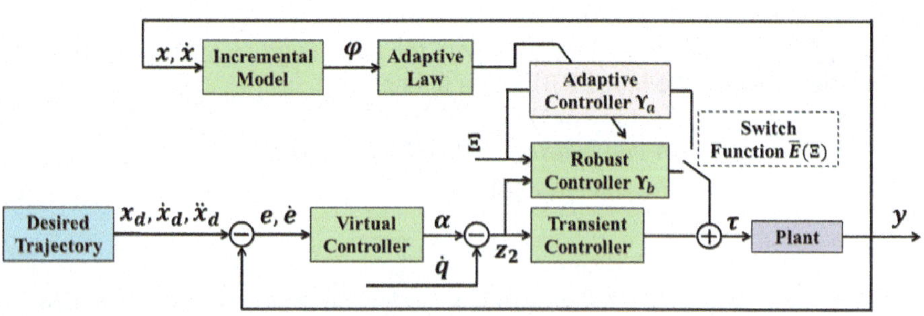

Fig. 4.2 Control diagram of adaptive neural controller using broad learning system

In general, the tracking error vector in the task space is designed as

$$\begin{aligned} z_1(t) &= x(t) - x_d(t) \\ z_2(t) &= \dot{q} - \alpha \end{aligned} \tag{4.14}$$

where $x_d(t) \in \Omega_d, \forall t \geq 0$ is the desired trajectory defined in task space, $\dot{q}$ stands for the joint velocity, α is the virtual controller which is defined as (4.20), z_1 denotes the tracking error in task space and z_2 represents the velocity tracking error of the manipulator in joint space.

In order to guarantee the prescribed tracking performance which can be achieved by evolving the tracking error $z_1(t)$ within a specified area, an exponential decaying performance function [28] is first considered as

$$\nu_i(t) = (\nu_{i0} - \nu_{i\infty}) \exp(-m_i t) + \nu_{i\infty} \tag{4.15}$$

where ν_{i0}, $\nu_{i\infty}$ and m_i, $i = 1, \cdots, N_d$ are the predefined positive constants. N_d is the manipulator's degree of freedom. Hence, the prescribed tracking performance can be described by the following inequalities:

$$-\beta_{a,i}\nu_i(t) < z_1(t) < \beta_{b,i}\nu_i(t) \tag{4.16}$$

where β_a and β_b are positive constants that can achieve the transient and steady-state performance integrated with ν_{i0}, $\nu_{i\infty}$ and m_i during the controller design.

4.3.2 Global NN Controller Design with Barrier Lyapunov Function and Backstepping

In this section, the principle of the controller design is given by the following form which is based on the backstepping method with an asymmetric barrier Lyapunov function (BLF) [29, 30]:

$$V_1 = \sum_{i=1}^{N_d} \left(\frac{h_i(z_1)}{2} \ln \frac{1}{1 - \delta_{b,i}^2} + \frac{1 - h_i(z_1)}{2} \ln \frac{1}{1 - \delta_{a,i}^2} \right) \tag{4.17}$$

where $\delta_{a,i} = \frac{z_1(i)}{\phi_{a,i}} = \frac{z_1(i)}{-\beta_{a,i}\nu_i}$, $\delta_{b,i} = \frac{z_1(i)}{\phi_{b,i}} = \frac{z_1(i)}{\beta_{b,i}\nu_i}$, and $h_i(z_1)$ is defined as

$$h_i(z_1) = \begin{cases} 1 & z_1 \geqslant 0 \\ 0 & otherwise \end{cases} \tag{4.18}$$

Then, the derivative of V_1 with respect to t is given by

$$\begin{aligned}\dot{V}_1 &= P^T \dot{z}_1 + \Pi \\ &= P^T (J(q)(z_2 + \alpha) - \dot{x}_d) + \Pi \end{aligned} \tag{4.19}$$

where $P = [p(1), p(2), \ldots, p(N_d)]$ with $p(i) = \frac{\chi_i^2}{(1-\chi_i^2)z_1(i)}$ denotes the transient control vector, $\Pi = \sum_{i=1}^{N_d} (\frac{(1-h_i)\delta_{a,i}^2}{(1-\delta_{a,i}^2)} \frac{\dot{\phi}_{a,i}}{\phi_{a,i}} + \frac{(h_i)\delta_{b,i}^2}{(1-\delta_{b,i}^2)} \frac{\dot{\phi}_{b,i}}{\phi_{b,i}})$. The virtual control law can be designed as

$$\alpha = J^{\dagger}(q)(\dot{x}_d - K_1 z_1 - L(t) z_1) \tag{4.20}$$

where $J^{\dagger} = J^T (J J^T)^{-1}$ denotes the Moore-Penrose inverse of $J(q)$, $K_1 = \text{diag}\,(k_{11}, k_{12}, \ldots, k_{1N_d})$ denotes the positive control vector. And $L(t) = \text{diag}(l_1(t), l_2(t), \ldots, l_{N_d}(t))$ with $l_i(t) = \sqrt{(\frac{\dot{\phi}_{a,i}}{\phi_{a,i}})^2 + (\frac{\dot{\phi}_{b,i}}{\phi_{b,i}})^2 + c_i}$, where c_i is also a positive constant to guarantee the boundedness of $\dot{\alpha}_i$ while $\dot{\phi}_{a,i}$ and $\dot{\phi}_{b,i}$ tend to be zero. Substituting (4.20) into (4.19) yields

$$\dot{V}_1 = P^T J(q) z_2 - P^T (K_1 z_1 + L(t) z_1) + \Pi \tag{4.21}$$

Considering the inequality described below:

$$l_i(t) - h_i \frac{\dot{\phi}_{a,i}}{\phi_{a,i}} - (1 - h_i) \frac{\dot{\phi}_{b,i}}{\phi_{b,i}} \geqslant 0 \tag{4.22}$$

it can be obtained that

$$\dot{V}_1 \leqslant - \sum_{i=1}^{N_d} k_1(i) \frac{\chi_i^2}{(1 - \chi_i^2)} + P^T J(q) z_2; \tag{4.23}$$

where $\chi_i = h_i \delta_{b,i} + (1 - h_i)\delta_{a,i}$. Next, we choose a positive definite Lyapunov function:

$$V_2 = V_1 + \frac{1}{2} z_2^T M(q) z_2 \tag{4.24}$$

In terms of (4.13), (4.14) and properties 1 and 2, the derivative of V_2 becomes

$$\begin{aligned}\dot{V}_2 &= \dot{V}_1 + \frac{1}{2} z_2^T \dot{M}(q) z_2 + z_2^T M(q) \dot{z}_2 \\ &= \dot{V}_1 + z_2^T (\tau - G(q) - C(q, \dot{q})\alpha - M(q)\dot{\alpha}) \\ &= \dot{V}_1 + z_2^T (\tau + F(\Xi)) \end{aligned} \tag{4.25}$$

where $\Xi = [q, \dot{q}, \alpha, \dot{\alpha}]^T$ and $F(\Xi)$ is an approximation function because the dynamic model of $M(q)$, $C(q, \dot{q})$ and $G(q)$ are unknown. Here, the incremental RBFNN (4.12) which has been discussed above is used to estimate the unknown function.

Inspired by the recently published work [14], the smooth switching function $\overline{E}(\Xi) = \text{diag}(E_1(\Xi), E_2(\Xi), \ldots, E_{N_d}(\Xi))$ is developed, where $E(\Xi) = \prod_{k=1}^{4N_d} e(\Xi(t))$ with the format of

$$e(\Xi(t)) = \begin{cases} 1 & |\Xi(t)| < d_a \\ \frac{d_b^2 - \Xi^2(t)}{d_b^2 - d_a^2} \exp(\frac{\Xi^2(t) - d_a^2}{\omega_i (d_b^2 - d_a^2)})^2 & otherwise \\ 0 & |\Xi(t)| > d_b \end{cases} \tag{4.26}$$

where $d_a = \varepsilon_a * q_f$, $d_b = \varepsilon_b * q_f$. Here q_f denotes the radius of the farthest neural node from the origin of the coordinates. ε_a and ε_b denote the distance threshold.

Hence, the global adaptive incremental NN controller is designed as

$$\tau_x = -K_2 z_2 - P^T J(q) - \overline{E}(\Xi)\Upsilon_a - (1 - \overline{E}(\Xi))\Upsilon_b \tag{4.27}$$

where K_2 is a positive definite vector, Υ_a, Υ_b are neural control terms and robust control terms respectively which are designed as

$$\Upsilon_a = \hat{Y}(\Xi) \tag{4.28}$$

$$\Upsilon_b = Y^U(\Xi) R \left(\frac{Y^U(\Xi) z_2}{\varpi} \right) \tag{4.29}$$

where $\hat{Y}$ is the approximation of (4.12), $Y^U(\Xi) = \text{diag}\{y_1^U(\Xi), y_2^U(\Xi), \ldots, y_{N_i}^U(\Xi)\}$. It is noted that $R(\frac{Y^U(\Xi)z_2}{\varpi}) = [\tanh(\frac{y_1^U(\Xi)z_2(1)}{\varpi(1)}), \tanh(\frac{y_2^U(\Xi)z_2(2)}{\varpi(2)}), \ldots, \tanh(\frac{y_{N_d}^U(\Xi)z_2(N_i)}{\varpi(N_i)})]^T$ where ϖ is a positive parameter.

Eventually, the learning law of incremental NN is designed as

$$\dot{\widehat{Wn}} = \Delta(\overline{E} z_2 Sn(\Xi) - \gamma \widehat{Wn}) \tag{4.30}$$

where Δ and γ are positive constant matrices.

Remark 4.2 Υ_a and Υ_b are two important parts in the incremental global adaptive NN controller (4.27). As shown in Fig. 4.3, the minimum and maximum compact subset Ω_a, Ω_b are variable in terms of the defined radius d_a and d_b which is dedicated to containing the desired compact set Ω_1. In such a case, it is worth noting that the input of broad RBFNN is always confined within the dynamic compact set such that a global tracking performance in the whole workspace can be achieved.

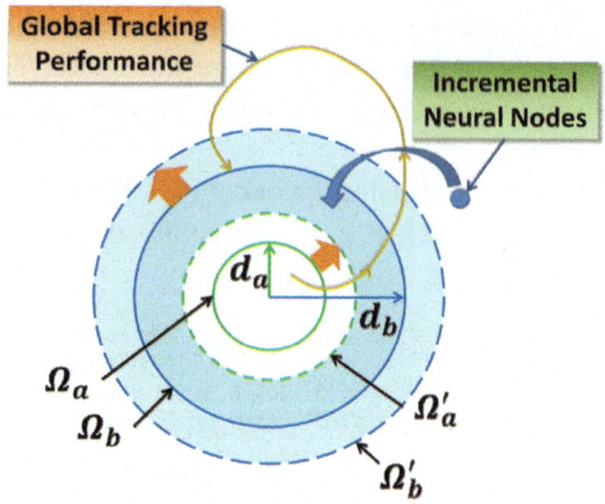

Fig. 4.3 The principle of incremental global NN control: solid cycles denote the initial compact subset and they will be changed to dash cycles when a new neural node is added to the network. The orange curve illustrates the process of global tracking performance. The NN controller Υ_a plays a leading role when the input vector lies in domain Ω_a. Instead, it will be displaced by the robust controller Υ_b once the input vector runs out of domain Ω_b and it will be forced to run back. If the input vector lies in between Ω_a and Ω_b, both controllers will work which are aimed to constrain the input run in the desired compact set Ω_1

4.3.3 Learning From Accumulated Knowledge

Lemma 4.1 *[31] By taking into account the PE condition [9] with the closed-loop control system (4.13), the control torque (4.27) and the incremental neural learning law (4.30), the neural estimated weights $\hat{W}$ can converge to the true or optimal weights W^* for any recurrent orbit $\Xi(t)$ with the satisfaction of PE and the approximation function (4.4) can be represented by $\hat{W}^T S(\Xi)$ or $\bar{W}^T S(\Xi)$:*

$$\begin{aligned} Y(\Xi) &= \hat{W}^T S(\Xi) + \epsilon(\Xi) \\ &= \bar{W}^T S(\Xi) + \bar{\epsilon}(\Xi) \end{aligned} \tag{4.31}$$

where $\bar{W} = mean_{t \in [t_a, t_b]} \hat{W}(t)$ represents the time segment after transient process with $t_b > t_a > T_0$, T_0 denotes a finite time.

In terms of Lemma 4.1, the training knowledge of the recurrent orbits $\Xi_i(t)$ can be stored in $\bar{W}_i$ which can be reused for similar control tasks even the new recurrent movements. Here, i denotes the training process of the ith action. This allows to achieve motor generalization via the accumulated knowledge. Specifically, the initial neural weights $\hat{W}_{i+1}(0)$ for the next new training process can be set as $\bar{W}_i$ as long as the manipulator works in the same workspace.

Remark 4.3 Compared with the conventional adaptive NN control system and deterministic learning control [6], the superiority of the proposed controller is that the accumulated knowledge can not only be reused for similar periodic orbits, but also be used to train other periodic orbits in the same workspace without retuning the controller parameters, which can enhance the extensibility and convenience in design of the controller.

4.4 Stability Analysis

In this part, the stability of a broad learning neural control system is rigorously established and the principle of how to determine the controller parameters in terms of Lyapunov theory is elaborated.

Theorem 4.1 *With the consideration of the manipulator system (4.13), combined with the filtered error model (4.14), control toque (4.27), neural weight learning law (4.30) and the incremental scheme (4.6), (4.7), for any periodic orbit x_d, whose initial condition is $x_d(0) \in \Omega_0$, the proposed learning scheme both at training stage or incremental stage can guarantee that all signals in the manipulator control system are uniformly ultimately bounded (UUB) by choosing the suitable design parameters of the system.*

Proof. Choose a Lyapunov function as

$$V = V_2 + \frac{1}{2}\sum_{i=1}^{N_i} \widetilde{Wn}_i^T \Theta^{-1} \widetilde{Wn}_i \tag{4.32}$$

Computing the derivative of (4.32) along (4.17), (4.24) and simplifying yield

$$\begin{aligned}
\dot{V} &= \dot{V}_2 + \sum_{i=1}^{N_i} \widetilde{Wn}_i^T \Theta^{-1} \dot{\widetilde{Wn}}_i \\
&= \dot{V}_1 + z_2^T(\tau + Y(\Xi)) + \sum_{i=1}^{N_i} \widetilde{Wn}_i^T \Theta^{-1} \dot{\widetilde{Wn}}_i \\
&= \dot{V}_1 - K_2 z_2^T z_2 - P^T J z_2^T \\
&\quad + \sum_{i=1}^{N_i} z_2(i)[-\overline{E}_i(\widehat{Wn}_i^T - Wn_i^{*T})Sn_i(\Xi) \\
&\quad + \overline{E}_i \widetilde{Wn}_i^T Sn_i(\Xi) + \epsilon(i)] \\
&\quad + \sum_{i=1}^{N_i} z_2(i)(1 - \overline{E}_i)(y_i - y_i^U(\Xi)\tanh(\frac{y_i^U(\Xi) z_2(i)}{\varpi(i)})) \\
&\quad + \sum_{i=1}^{N_i} [-\gamma(i) \widetilde{Wn}_i^T (Wn_i^* + \widetilde{Wn}_i)]
\end{aligned} \tag{4.33}$$

And it is easy to have the following inequalities in terms of the Young's inequality:

$$y_i z_2(i) - y_i^U z_2(i) \tanh\left(\frac{y_i^U z_2(i)}{\varpi(i)}\right) \leqslant \iota\varpi(i) \tag{4.34}$$

$$\begin{aligned} \dot{V} \leqslant & \sum_{i=1}^{N_d}\left[-K_1(i)\frac{\chi_i^2}{(1-\chi_i^2)}\right] \\ & + \sum_{i=1}^{N_i}\left[-(K_2(i)-\frac{1}{2})z_2^2(i) - \frac{1}{2}\gamma(i)\left\| \begin{matrix}\tilde{W}_i \\ \tilde{W}_{exi}\end{matrix} \right\|^2\right] \\ & + \sum_{i=1}^{N_i}\left[\frac{1}{2}\gamma(i)\left\| \begin{matrix} W_i^* \\ W_{exi}^* \end{matrix}\right\|^2 + \frac{1}{2}\epsilon^2(i) + \iota\varpi(i)\right] \end{aligned} \tag{4.35}$$

Considering the inequality

$$-\frac{\chi_i^2}{(1-\chi_i^2)} \leqslant -\ln\frac{1}{(1-\chi_i^2)} \quad \forall|\chi_i| < 1 \tag{4.36}$$

it can be obtained that

$$\begin{aligned} \dot{V} \leqslant & \sum_{i=1}^{N_d}\left[-K_1(i)\ln\frac{1}{(1-\chi_i^2)}\right] \\ & + \sum_{i=1}^{N_i}\left[-K_3(i)z_2^2(i) - \frac{1}{2}\gamma(i)\left\| \begin{matrix}\tilde{W}_i \\ \tilde{W}_{exi}\end{matrix} \right\|^2\right] \\ & + \sum_{i=1}^{N_i}\left[\frac{1}{2}\gamma(i)\left\| \begin{matrix} W_i^* \\ W_{exi}^* \end{matrix}\right\|^2 + \frac{1}{2}\epsilon^2(i) + \iota\varpi(i)\right] \end{aligned} \tag{4.37}$$

where $K_3 = K_2 - \frac{1}{2}$. In accordance with (4.17), (4.24) and (4.32), it follows that

$$V = Q + \frac{1}{2}z_2^T M(q) z_2 + \frac{1}{2}\sum_{i=1}^{N_i} \tilde{W}_i^T \Theta^{-1} \tilde{W}_i \tag{4.38}$$

where $Q = \sum_{i=1}^{N_d}(\ln\frac{1}{1-\chi_i^2})$. Hence, inequality (4.37) can be rewritten as

$$\dot{V}(t) = \eta_s V(t) + \mu_s \tag{4.39}$$

where $\eta_s = \min\{\lambda_{min}(K_1), \frac{2\lambda_{min}(K_3)}{\lambda(M(q))}, \frac{\gamma}{\lambda_{max}(\Theta^{-1})}\}, \mu_s = \sum_{i=1}^{N_i}(\frac{1}{2} \parallel W_i^* \parallel^2 + \frac{1}{2}\epsilon^2(i) + \iota\varpi(i))$. Assume that $\beta_s = \frac{\mu_s}{\eta_s}$ and it follows from (4.38) and (4.39) that

$$V(t) \leqslant (V(0) - \beta_s)exp(-\eta_s t) + \beta_s \leqslant V(0) + \beta_s \tag{4.40}$$

It can be concluded from (4.32) and (4.39) that $\dot{V}$ is definitely negative. Thus it can be obtained that the filter error z_2 as well as the weight approximation errors $\hat{W}$ are bounded. According to the definition 4.15 and 4.16, it can be also concluded that ϕ_a and ϕ_b are bounded which implies that the tracking error z_1 is bounded and the predefine tracking performance is satisfied. Hence the states x and $J(q)$ are bounded. Consequently, all the signals in the manipulator control system remain UUB during the training and incremental stage. In this way, the system is stable in the sense of Lyapunov both in the learning stage and incremental stage. This completes the proof. □

4.5 Simulation

In order to test the validness of the developed broad learning enabled adaptive neural control method with application on modeling motor skills generalization, a set of simulations with the same scenarios studied in [4] is performed under the external force field by utilizing a two-joint manipulator as shown in Fig. 4.4. The parameters of the manipulator are designed as the mass of links $m_1 = m_2 = 10.0kg$, length of links $l_1 = l_2 = 1.0m$, inertia of links $I_1 = I_2 = 0.83kgm^2$ and gravity acceleration $g = 9.8m/s^2$.

During the training process, a set of basic training actions with four straight line movements spanning a range of 360° in an ellipse is defined. After training the predefined movements, the learning knowledge will be used to control the manipulator to draw the ellipse which is defined as $\frac{(x-1)^2}{0.09} + \frac{y^2}{0.04} = 1$ in the same workspace. In this simulation, the parameters of the predefined tracking performance are set to $\nu_{i0} = [1.25, 1.25]$, $\nu_{i\infty} = [0.1, 0.1]$, $\beta_a = \beta_b = 0.2$. The control torque is designed as 4.27 with $K_1 = \text{diag}(30, 30)$, $K_2 = \text{diag}(30, 30)$, $\overline{\varpi} = [0.1, 0.12]$. The internal force vector is specified as $f_d = [0, 2]$.

Aiming at providing convincing comparison results, RBFNN, FNN [32–34] and the proposed method are used to complete the same fundamental skills training

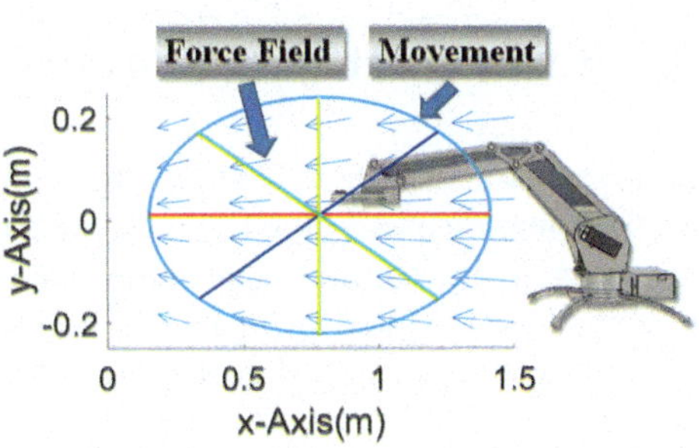

Fig. 4.4 Simulation setup for broad learning enabled adaptive neural control: a 2-DOF manipulator is required to complete the reaching and cycle-drawing movements under the force field disturbance

and knowledge reusing with the same parameters set designed above. Furthermore, $2^8 = 256$ nodes are specified for both RBFNN and FNN whose centers are located at the space of $\Phi_1 = [-1, 1] \times [-1, 1] \times [-1, 1] \times [-1, 1] \times [-1, 1] \times [-1, 1] \times [-1, 1] \times [-1, 1]$. The switching radius d_a and d_b will be constant with 5 and 20 respectively in this case. Instead of specifying the neuron centers beforehand which are notoriously hard to design approximately, the proposed method only needs to design the center distance threshold Θ which is specified as $\Theta = 20, 30, 50$ in this simulation.

The simulation results of tracking performance are illustrated in Fig. 4.5. It can be seen that the overall tracking by using the proposed method is superior to the conventional adaptive control. Moreover, the tracking error z_1 corresponding to the RBFNN, FNN and proposed method with different distance thresholds is shown in Fig. 4.6. From these simulations, it can be objectively concluded that applying the broad learning enabled adaptive NN control (4.27) with the incremental adaptive law (4.30), the manipulator can draw the ellipse with higher accuracy and fewer oscillations in use of the accumulated knowledge. Notably, the predetermined distance threshold Θ is the most important parameter to affect the structure of the neural network since it determines whether the new node ought to be added or not. It is clear that the suitable distance Θ it uses, the tracking error can become smaller whereas it will lead to more oscillatory at the early stage.

In the following, Figs. 4.8, 4.9 and 4.10 show the simulation results for partial weights convergence of the incremental neural network with the corresponding distance threshold Θ. The compared results show that only partial weights converge to relatively large values, whereas others remain a small neighborhood of zero. This is because the large estimated weights of the incremental neural network, whose center points are near the input orbit, can be activated and will converge to the true value W^*. However, the neuron weights that are far away from the input vector are not activated and will stay at a small range of zero. Obviously, these results are consistent with the requirement of the partial PE condition.

In addition, the number of the used neural nodes in incremental neural networks during the convergence to optimal neural weights is shown in Fig. 4.7. In Fig. 4.7, it is clear that the number of neurons grows fast with the decreasing of the distance threshold Θ.

4.6 Experiment and Discussion

In this section, the effectiveness of the proposed control method will be evaluated by implementation on a 7-DoF Baxter robot arm which provides a variety of interfaces to get access to the real-time state of the manipulator through a standard robot operating system (ROS). Moreover, it is also convenient for researchers to utilize different control strategies, i.e., position, velocity or torque control, to perform specific tasks.

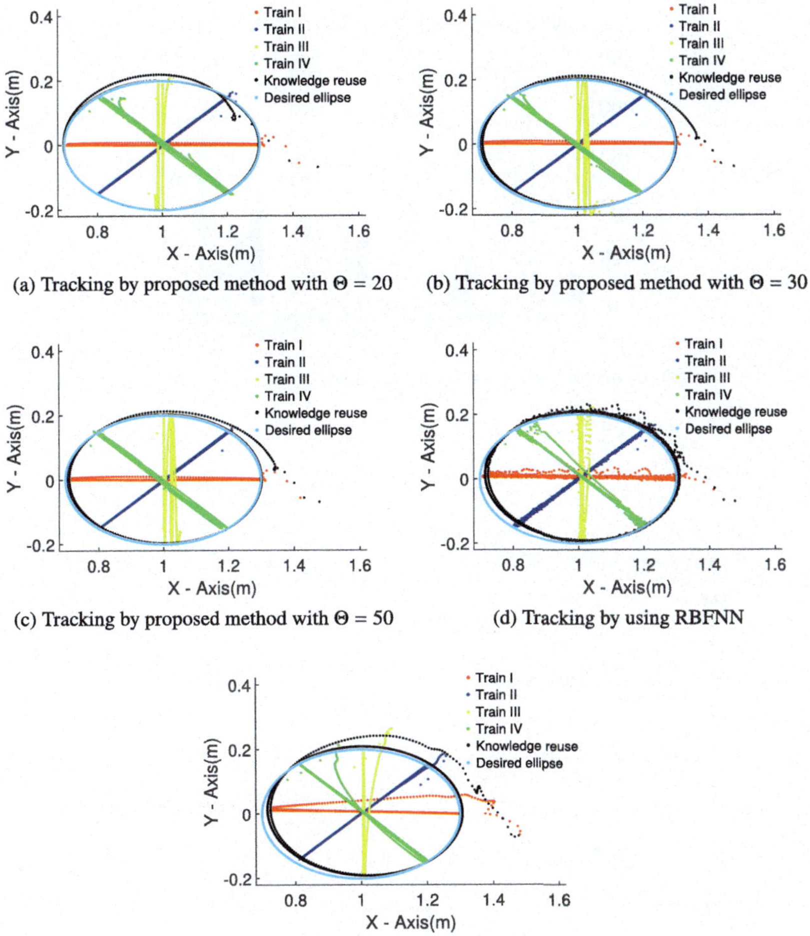

(a) Tracking by proposed method with Θ = 20 (b) Tracking by proposed method with Θ = 30

(c) Tracking by proposed method with Θ = 50 (d) Tracking by using RBFNN

(e) Tracking by using FNN

Fig. 4.5 Simulation results of trajectory tracking for proposed algorithm and knowledge reusing

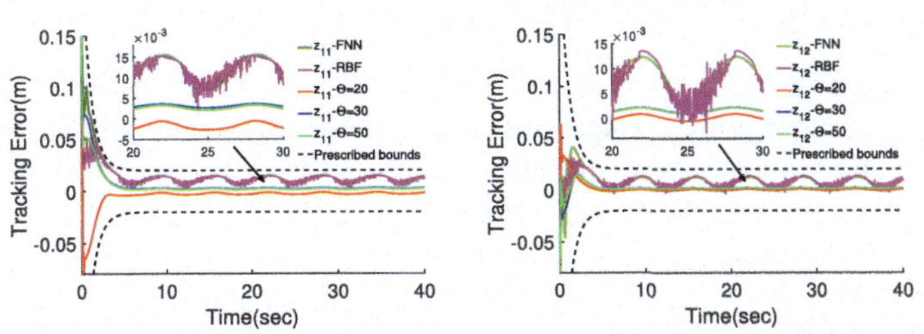

(a) Comparison of tracking error for X-axis (b) Comparison of tracking error for Y-axis

Fig. 4.6 Comparison of tracking error with $\Theta = 20, 30, 50$ and traditional RBFNN respectively

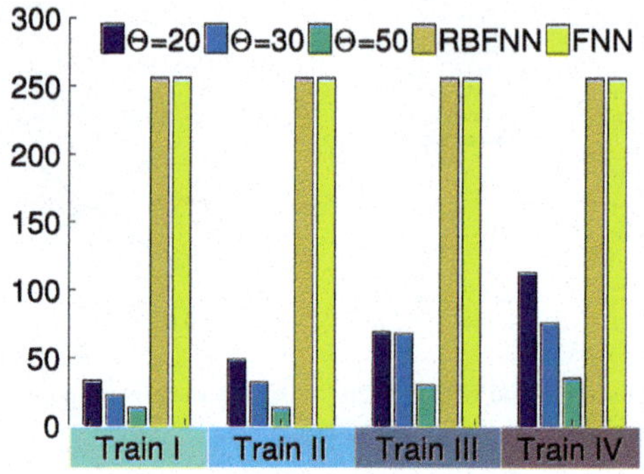

Fig. 4.7 The number of neural nodes in proposed method

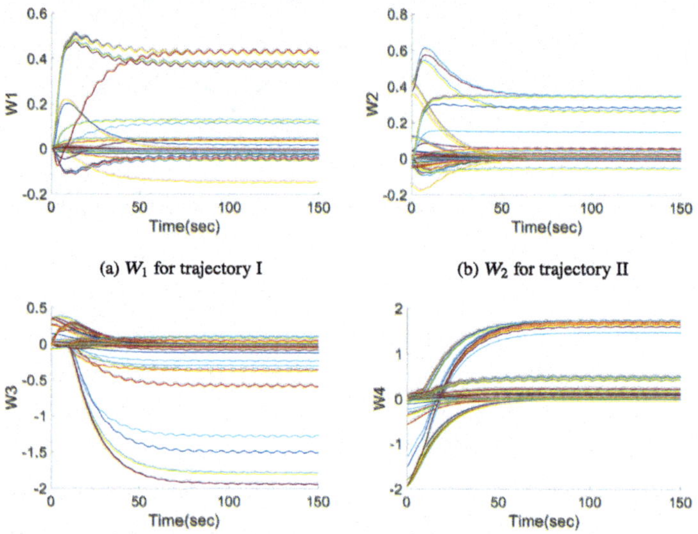

(a) W_1 for trajectory I

(b) W_2 for trajectory II

(c) W_3 for trajectory III

(d) W_4 for trajectory IV

Fig. 4.8 Partial weights convergence with $\Theta = 20$

In the experiment, only a single arm is employed. Moreover, in order to denote the payload and simulate the disturbance, an object with 1.5 kg in weight has been attached but not fixed on the end-effector. As shown in Fig. 4.11, two phases (phase I and phase II) are defined, in which the robot arm is trained to pick and to place an object in a different space. Then the knowledge, which is learned through training as embedded into the RBFNN, will be reused to complete the task of phase III. Considering the approximation accuracy and computational efficiency, the input of NN is defined as $[q^T, \dot{\alpha}^T] \in \mathbb{R}^{14}$. In order to well compare the performance between RBFNN and the proposed algorithm, the fixed neural nodes of RBFNN are first specified by trial and error as $[0.35, 0.55] \times [-0.75, -0.6] \times [0.75, 0.8] \times [0.8, 0.1] \times$

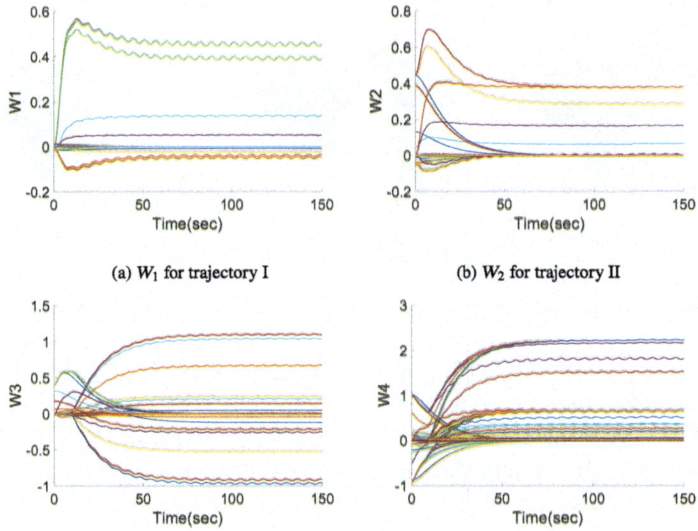

(a) W_1 for trajectory I

(b) W_2 for trajectory II

(c) W_3 for trajectory III

(d) W_4 for trajectory IV

Fig. 4.9 Partial weights convergence with $\Theta = 30$

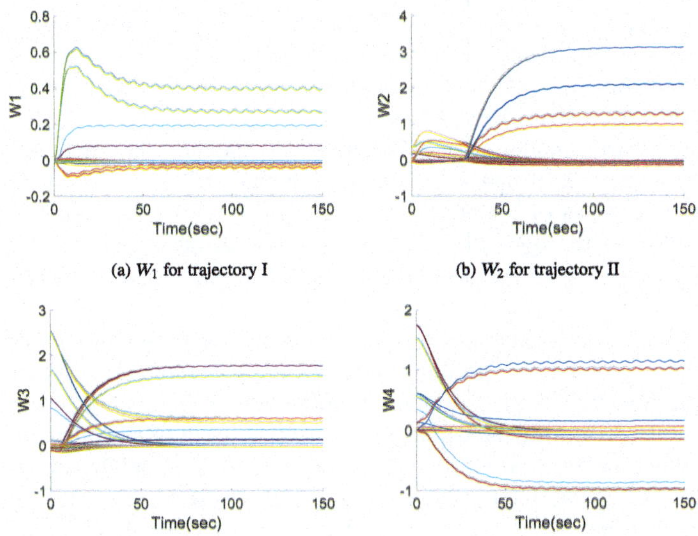

(a) W_1 for trajectory I

(b) W_2 for trajectory II

(c) W_3 for trajectory III

(d) W_4 for trajectory IV

Fig. 4.10 Partial weights convergence with $\Theta = 50$

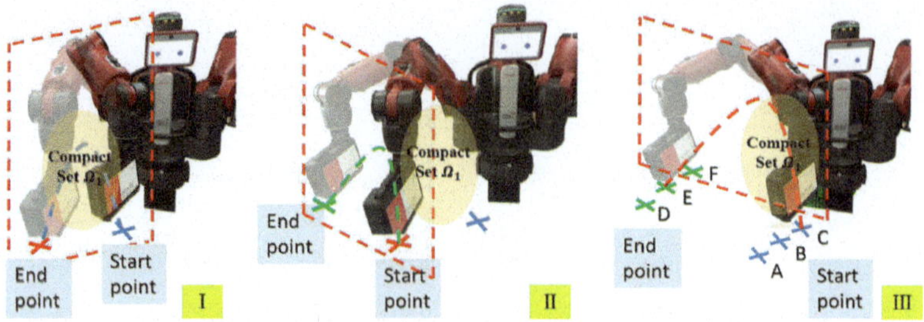

Fig. 4.11 Experimental setup during phase I–III for motor training and generalizing

$[-0.55, -0.45] \times [1.35, 1.55] \times [-0.02, 0.4] \times [0.45, 0.65] \times [-0.5, -0.25] \times [0.35, 0.5] \times [0.7, 0.9] \times [-0.2, 0.15] \times [0.85, 1.05] \times [0.15, 0.5]$ with totally $2^{14} = 16384$ nodes. Such centers of neural networks are chosen between the upper and lower bound of motion range and speed limits of each joint in the phase I which assumes that they are distributed in the compact set Ω_1 approximately as shown in Fig. 4.11.

On the other hand, the proposed method is required to complete a set of tasks with the training knowledge in phase III. In this part, three start points A, B, C are firstly set as $[0.7, -0.2 - 0.25]$, $[0.65, -0.15 - 0.25]$, $[0.6, -0.1 - 0.25]$ and three end points D, E, F as $[0.5, -0.8, 0.25]$, $[0.45, -0.75, 0.25]$, $[0.4, -0.7, 0.25]$ in Cartesian space. And then a set of tasks containing 9 trajectories can be obtained by permuting the start points and end points. Equally, the initial position of the manipulator is located at $[0.58, -0.26, 0.1]$.

For the control model, the parameters K_1 and K_2 are defined as $K_1 = \text{diag}\{10, 8, 11.5, 0, 0, 0\}$ and $K_2 = \text{diag}\{5.5, 5, 6, 4, 8.5, 9, 5\}$. The parameters of performance function (4.15) are set as $\nu_{i0} = [1.05, 1.05, 1.05, 0, 0, 0]$ and $\nu_{i\infty} = [0.5, 0.5, 0.5, 0, 0, 0]$, $\beta_a = \beta_b = 0.2$. The switch radius d_a and d_b in (4.26) are remained constant with 4 and 9. Also, the predetermined distance threshold Θ is set as 1.

The tracking error of the 9 trajectories by using the proposed method in phase III is shown in Fig. 4.12. From the results, it can be concluded that the predefined tracking performance that has used the proposed controller can be achieved, while obvious oscillatory during the whole process occurs for the RBFNN controller. This is because the training experience in phase II, which is far from the compact set Ω_1, may bring less effect to the knowledge reuse at phase III. Furthermore, it should be emphasized that the number of neural nodes that are incremented during the broad learning in RBF has been reduced sharply with 58 nodes in phase I and 72 nodes in phase II respectively.

Fig. 4.12 Tracking error of proposed algorithm and RBFNN control during knowledge reusing

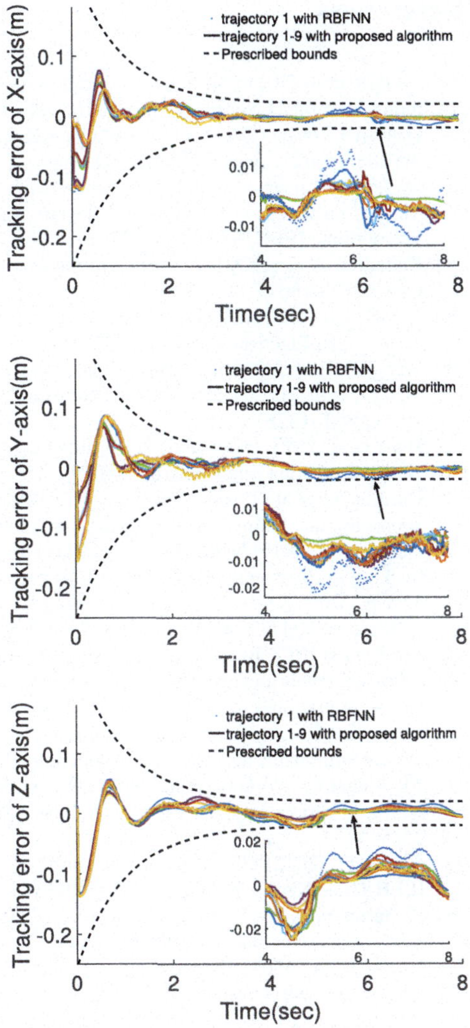

4.7 Conclusion

In this chapter, a novel neural adaptive control using a broad learning system based on deterministic learning is proposed. The incremental RBF neural network, which can adjust and augment the neural node dynamically, is studied to estimate the unknown model of the manipulator dynamic. During the training and knowledge reusing stage, the tracking error can satisfy the predefined tracking performance. It has been proven that the closed-loop system can be remained globally stable by using the incremental adaptive controller. Simulation and experiment results have shown that the proposed framework is valid for motor generalizing learning and can use the accumulated knowledge to fulfill new tasks under the dynamic external disturbance environment.

References

1. Jj ODX et al (2011) Stimulation of the human motor cortex alters generalization patterns of motor learning. J Neurosci 31(19):7102–7110
2. Ganesh G et al (2017) Biomimetic motor behavior for simultaneous adaptation of force, impedance and trajectory in interaction tasks. In: Paper presented at the 2010 IEEE international conference on robotics and automation, Anchorage, AK, USA, 03-07 May 2010
3. Yang C et al (2011) Human-like adaptation of force and impedance in stable and unstable interactions. IEEE Trans Robot 27(5):918–930
4. Kadiallah A et al (2012) Generalization in adaptation to stable and unstable dynamics. Plos One 7(10):e45075
5. Ge SS et al (1997) Adaptive neural network control of robot manipulators in task space. IEEE Trans Indus Electron 44(6):746–752
6. Wang C, Hill DJ (2006) Learning from neural control. IEEE Trans Neural Netw 17(1):130–146
7. Wang C et al (2009) Deterministic learning of nonlinear dynamical systems. Int J Bifurc Chaos 19(4):1307–1328
8. Wu Y et al (2013) Deterministic learning based adaptive network control of robot in task space. Acta Automatica Sinca 39(6):806–815
9. Kurdila AJ et al (1995) Persistency of excitation in identification using radial basis function approximants. Siam J Control Optim 33(2):625–642
10. Dai SL et al (2015) Neural learning control of marine surface vessels with guaranteed transient tracking performance. IEEE Trans Indus Electron 63(3):1717–1727
11. Min W, Yang A (2017) Dynamic learning from adaptive neural control of robot manipulators with prescribed performance. IEEE Trans Syst Man Cybern Syst 47(8):2244–2255
12. Liu Z et al (2017) Broad learning system: feature extraction based on K-means clustering algorithm. In: Paper presented at the 2017 4th international conference on information, cybernetics and computational social systems (ICCSS), Dalian, China, 24–26 Jul 2017
13. Chen CLP, Liu Z (2017) Broad learning system: an effective and efficient incremental learning system without the need for deep architecture. IEEE Trans Neural Netw Learn Syst 29(1):10–24
14. Xu B et al (2015) Global neural dynamic surface tracking control of strict-feedback systems with application to hypersonic flight vehicle. IEEE Trans Neural Netw Learn Syst 26(10):2563–2575
15. Huang JT (2012) Global tracking control of strict-feedback systems using neural networks. IEEE Trans Neural Netw Learn Syst 23(11):1714–1725
16. Kostarigka AK, Doulgeri Z, Rovithakis GA (2013) Prescribed performance tracking for flexible joint robots with unknown dynamics and variable elasticity. Automatica 49(5):1137–1147
17. Park J et al (2005) Direct adaptive controller for nonaffine nonlinear systems using self-structuring neural networks. IEEE Trans Neural Netw 16(2):414–422
18. Barakat M et al (2011) Self adaptive growing neural network classifier for faults detection and diagnosis. Neurocomputing 74(18):3865–3876
19. Jia Z, Song Y (2016) Barrier function-based neural adaptive control with locally weighted learning and finite neuron self-growing strategy. IEEE Trans Neural Netw Learn Syst 28(6):1439–1451
20. Pao YH et al (1994) Learning and generalization characteristics of the random vector functional-link net. Neurocomputing 6(2):163–180
21. Park J, Sandberg IW (1991) Universal approximation using radial-basis-function networks. Neural Comput 3(2):246–257
22. Conditt MA et al (1997) The motor system does not learn the dynamics of the arm by rote memorization of past experience. J Neurophysiol 78(1):554–560
23. Karim HT et al (2017) Motor sequence learning-induced neural efficiency in functional brain connectivity. Behav Brain Res 319:87–95
24. Willingham DB (1998) A neuropsychological theory of motor skill learning. Psychol Rev 105(3):558
25. Shadmehr R, Mussa-Ivaldi FA (1994) Adaptive representation of dynamics during learning of a motor task. J Neurosci 14(5):3208–3224
26. Patra JC, Sandberg IW (1999) Identification of nonlinear dynamic systems using functional link artificial neural networks. IEEE Trans Syst Man Cybern 29(2):254–262

27. Ge SS, Harris CJ (1998) Adaptive neural network control of robotic manipulators. World Scientific Publishing Company, Singapore
28. Bechlioulis CP, Rovithakis GA (2008) Robust adaptive control of feedback linearizable MIMO nonlinear systems with prescribed performance. IEEE Trans Autom Control 53(9):2090–2099
29. Tee KP et al (2011) Control of nonlinear systems with time-varying output constraints. Automatica 47(11):2511–2516
30. Tee KP et al (2009) Barrier Lyapunov Functions for the control of output-constrained nonlinear systems. Automatica 45(4):918–927
31. Dai SL et al (2012) Identification and learning control of ocean surface ship using neural networks. IEEE Trans Indus Inform 8(4):801–810
32. Liu YJ et al (2016) Fuzzy approximation-based adaptive backstepping optimal control for a class of nonlinear discrete-time systems with dead-zone. IEEE Trans Fuzzy Syst 24(1):1–1
33. Wai RJ, Muthusamy R (2013) Design of fuzzy-neural-network-inherited backstepping control for robot manipulator including actuator dynamics. IEEE Trans Fuzzy Syst 22(4):709–722
34. He W, Dong Y (2017) Adaptive fuzzy neural network control for a constrained robot using impedance learning. IEEE Trans Neural Netw Learn Syst 29(4):1174–1186

Chapter 5
Hybrid Learning and Control Using Improved Dynamical Movement Primitive and Adaptive Neural Network Control

5.1 Introduction

In recent decades, learning from demonstrations (LfD), a technique that develops strategies from example states to action mappings [1], has attracted considerable attention along with the development of robotics and AI technologies. A recent survey on LfD concluded that current limitations of LfD include representation of complex behaviours, reliance on labelled data, and suboptimal and inappropriate demonstrators [2]. To solve this problem, this section proposes an incremental skill learning and generalization framework to enable robots to modify simple initial actions to complex cases. This method is based on motion primitive (MP) technology, in which a long-term complex motion is divided into multiple sub-actions. Then, the sub-actions are extracted into MPs and finally these MPs are reprogrammed and generalised to fit a new task [2].

The MP can be presented in many forms, e.g. Kernelized Movement Primitive (KMP) [3], Compliant Movement Primitive (CMP) [4] and Dynamical Movement Primitive (DMP) [5]. DMP was proposed by Ijspeert et al. [6, 7] and then improved by many researchers. In addition to the classical DMP, there are a number of improved methods such as discrete DMP, periodic DMP, etc., and some scholars combined DMP with reinforcement learning (RL) [8, 9], deep learning [10], life-long learning and various control methods [11–13] to expand the scope of DMP. DMP has a very concise expressions that is a second-order function with only three variables and a forcing function. And the applications of DMP contain trajectory tracking in Euclidean space [14], EMGs signal prediction [26], force control in a contact manipulation [16], motion and state monitoring [17] and special tasks such as obstacle avoidance [15, 18], cooperative manipulations [19, 20] and multi-modal skill learning.

The limitation of the classical DMP is that once the skills are learned, the characteristics expressed by the forcing function are fixed. Even though some variables, e.g. position, velocity, can be generalized in space and time by modifying the starts, goals

C. Yang et al., *Advanced Teleoperation and Robot Learning for Dexterous Manipulation*,
Springer Tracts in Advanced Robotics 160,
https://doi.org/10.1007/978-3-031-78501-6_5

and scaling factors. Some improvements of DMP in [15, 18–20] are made by adding additional terms for obstacle avoidance and cooperative manipulation. However, the terms are specially designed by using time-related variables such as position, angle, and velocity, etc., which cannot be generalized in phase space, like 's' in the forcing function. If operational requirements keep changing, the newly learned skills and added terms should update, which costs a lot of time and increases the complexity of computation. Reinforcement learning is used to achieve DMP-based incremental skill learning. For example, Matteo et al. proposed an incremental point-to-point motions learning method based on a dynamical system. For a new demonstration, the original dynamical system will be redesigned to approach the new task [21]. Lemme proposed a bootstrapping cycle to build a suitable primitive library [22]. In this library, the old primitives are refined and new ones are added, while the unused ones are deleted. Yuan et al. [8] and Li et al. [9] followed the similar technique and updated weights by integrating probability-weighted RL to realize skill modification. Wang and Wu et al. [23, 24] proposed DMP plus (DMP+) method to realize efficient skill modifications by using truncated kernels and local biases to achieve two contributions. One is preserving the desirable properties of the original skill and achieving lower mean square errors (MSEs). The other is the reusability of existing primitives, which can reduce human fatigue in imitation learning and correcting errors in demonstration without requiring further demonstration. Compared with RL-based methods, DMP+ requires less computation and retains the original features, which is used as a benchmark method for comparison in this method.

The combination of DMP and robot control is another topic that attracts much attention. Schaal et al. [6, 25], proposed a framework for motor control combining DMP. In our previous research, we combined DMP and adaptive NN control [26], admittance control [27], and neural networks [28–31] for robot control. In this section, inspired by DMP+, we propose a novel incremental skill learning and generalization method called incremental DMP (IDMP). The forcing function of IDMP can be incrementally updated by adding new features and weights to track new trajectories. Considering uncertain dynamics parameters and state tracking limitations, an adaptive NN controller is designed, in which the NN term is also incrementally updated and can accumulate and reuse the learned knowledge to improve the learning efficiency. System stability are ensured by building a barrier Lyapunov function (BLF). The proposed framework is shown in Fig. 5.1: First, an old trajectory is expressed by DMP with a forcing function. After linear transformation of the feature vector, we can obtain the incremental term to compose a IDMP function to achieve a new trajectory, which is then transferred by an inverse kinematic solver to obtain the joint information as the input of the control part. The real-time joint tracking errors are applied to create a virtual controller by BLF. We consider dynamics uncertainties and contact force estimation errors and use adaptive control term to estimate and compensate the errors based on incremental adaptive NN control to ensure system stability. Compared to DMP-based trajectory planning and various control methods, the proposed framework offers three advantages:

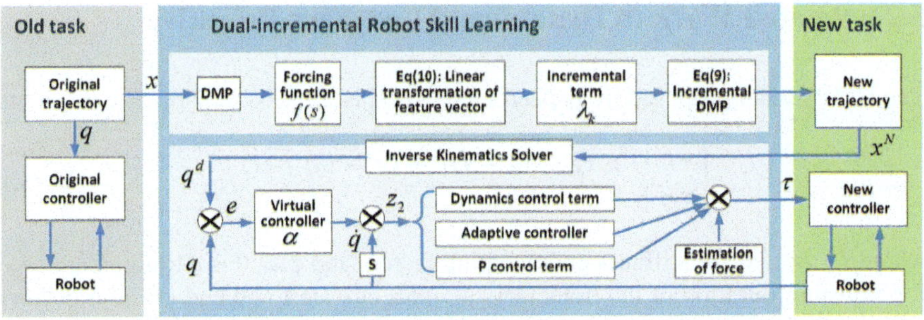

Fig. 5.1 Diagram of trajectory and force dual-incremental robot skill learning and generalization

(a) **Skill incremental learning and original skill preservation**

Similar to DMP+ and Acnmp [32], the old skills can be preserved during the gradual adaptation process to new situations, so they can be easily recovered for the old situations. The difference in the computation from DMP+ is the skill adaptation is realized by adding new linear transformations of the existing kernels rather than changing kernels, which gives the forcing function with a stronger nonlinear adaptability and makes it more suitable for the dynamic skill learning process without adding new kernels.

(b) **High-accuracy trajectory tracking and multi-style skill transformation**

According to the board learning in [33], the preliminary NN can achieve better performance after inserting additional extension nodes, which have a similar function as the linear transformations of IDMP. Therefore, we use IDMP to improve trajectory tracking performance and realize multi-style skill transformation. Multi-style skill transformation suits the situation that the original skill is ambiguous and leads to different styles in motion sequence [34]. An example, like the following second experiment, is a letter recognized as an 'a' first, and it is probably a 'u' or a 'v' after confirmation. We show the skill transformation process from one shape to multi-stylistic shapes based on the same original and extended features, but with different weights.

(c) **Adaptive NN control with constraints on the transient tracking errors**

After renewing trajectory using IDMP, robot system controller should be improved to minimize tracking errors to the updated trajectory. Additionally, the controller should process the uncertain dynamics parameters, force estimation errors, and limitations on the transient state errors. In this section, we proposed an adaptive NN and BLF-based controller, where the NN nonlinear fitting part can increase the number of neurons and update the weights, so that it can accumulate and reuse the learned weights to improve the learning efficiency.

5.2 Related Work to Dynamical Movement Primitive

The DMP model proposed by Ijspeert et al. [6, 7] is

$$\begin{cases} \tau \dot{v} = K(g - x) - Dv + (g - x_0) f(s) \\ \tau \dot{x} = v \end{cases}, \tag{5.1}$$

where $K, D > 0$ are stiffness and damping factors and $\tau > 0$ is a timing parameter for adjusting duration of the trajectory, x_0 and g are start and end of the trajectory. $f(s) = \theta^T \Psi(s)$ is a linear combination of the normalized Gaussian functions ψ_i, where ψ_i, ψ_i and w_i is the weight of ψ_i and the Gaussian functions ψ_i is expressed as

$$\psi_i = \frac{\varphi_i(s)s}{\sum_{i=1}^{n} \varphi_i(s)}, \varphi_i(s) = \exp(-h_i(s - c_i)^2), \tag{5.2}$$

where c_i and $h_i > 0$ are the centre and width of the radial basis function $\varphi_i(s)$. The transformation function (or forcing function) $f(s)$ is expressed by the phase variable s and a canonical system

$$\tau \dot{s} = -os, \qquad o > 0 \tag{5.3}$$

The converging time is modified by factor o to ensure $s \to 0$ at the end state for erasing the influence of $f(s)$ in (5.1). The θ is estimated by minimizing the function

$$\min \left(\sum_{k=1}^{N} \left(f_k^{Tar} - f(s) \right)^2 \right), \tag{5.4}$$

where $f^{Tar}(s)$ the target value of $f(s)$ that is calculated by the kth, $k = 1, 2, \ldots,$ K demonstrated trajectory x_d^k and velocity v_d^k:

$$f_k^{Tar} = \left(\tau \dot{v}_d - K\left(g - x_d^k\right) - Dv_d^k \right) / (g - x_0). \tag{5.5}$$

Remark 5.1 DMP method is also applied multi-style skill learning from multi-demonstrations, named Stylistic DMP (SDMP) [34]. The SDMP modifies (5-1) into

$$\begin{cases} \tau \dot{v} = K(g - x) - Dv + (g - x_0) \tilde{f}(s) \\ \tau \dot{x} = v \end{cases}, \tag{5.6}$$

where $\tilde{f}(s)$ is then expressed as $\tilde{f}(\mathbf{s}) = \sum_{j=1}^{J} f_j(s_j)$, $f_j(s_j) = \theta_j{}^T \Psi_j(s_j)$, $\theta_j = \left[w_{j1}, w_{j2}, \ldots, w_{jn}\right]^T$, $\Psi_j(s_j) = \left[\psi_{j1}, \psi_{j2}, \ldots, \psi_{jn}\right]^T$, $\psi_{ji} = \frac{\varphi_{ji}(s_j)s_j}{\sum_{i=1}^{n} \varphi_{ji}(s_j)}$, $\varphi_{ji}(s_j)$ $= \exp(-h_{ji}(s_j - c_{ji})^2)$. Set $\mathbf{s} = [s_1, s_2, \ldots, s_J]$ as a style parameter vector and

$\Theta = [\theta_1, \theta_2, \ldots, \theta_J]$ as a parameter matrix, the calculation purpose is to acquire the optimal parameter vector $\Theta* = [\theta_1*, \theta_2*, \ldots, \theta_J*]$.

Considering different skills expressed by DMP have the same expression as in (5.1), and the main difference focuses on the forcing function, we can update the to realize skill incremental learning. Following the idea of board learning system (BLS) [33] and [36], an efficient incremental learning system without the need for deep architecture, incremental learning algorithm has a promising performance in calculation accuracy and learning speed. Especially, with the increase of the enhancement nodes, the network can approach a nonlinear function with any accuracy, which inspires us to build an incremental updating forcing function that the vectors and can be extended to change the learned skills and fit new trajectories.

The main challenge is how to reshape θ and $\Psi(s)$ to enable the new added features and extended terms to satisfy the properties of DMP. For example, in (5.2), the feature variables $\psi_i, i = 1, 2, \ldots, m$ are normalized and satisfy

$$
\begin{aligned}
\sum_{i=1}^{n} \psi_i &= \frac{\varphi_1(s)s}{\sum_{i=1}^{n} \varphi_i(s)} + \frac{\varphi_2(s)s}{\sum_{i=1}^{n} \varphi_i(s)} + \cdots + \frac{\varphi_n(s)s}{\sum_{i=1}^{n} \varphi_i(s)} \\
&= \frac{\sum_{i=1}^{n} \varphi_i(s)s}{\sum_{i=1}^{n} \varphi_i(s)} \\
&= s.
\end{aligned}
\tag{5.7}
$$

If we set an extended function as $\Phi_j, j = 1, 2, \ldots, m$ and $\hat{\psi}_i$ as the modified term to $\psi_i, i = 1, 2, \ldots, n$, they will be normalized and satisfy:

$$
\sum_{i=1}^{n} \hat{\psi}_i + \sum_{j=1}^{m} \Phi_j = s. \tag{5.8}
$$

Then the main question in IDMP is how to generate Φ_j and modify $\hat{\psi}_i$ to enable (5.8) is satisfied. A lemma is presented for the following deduction process.

Lemma 5.1 *For matrices $A \in R^{n\times m}$, $W \in R^{m\times 1}$ and $Y \in R^{n\times 1}$ satisfying $Y = AW$, if A is extended to $\bar{A} = [A\,|a\,]$ $\bar{A} = [A\,|a\,]$, and $a \in R^{n\times k}$, then the new weight vector $\bar{W} \in R^{(m+k)\times 1}$ is calculated based on the W as*

$$
\bar{W} = \begin{bmatrix} W - db^T Y \\ b^T Y \end{bmatrix}, \tag{5.9}
$$

where $d = (A)^+ a$, $b^T = \begin{cases} c^+ & if\ c \neq 0 \\ (1 + d^T d)^{-1} d^T (A)^+ & if\ c = 0 \end{cases}$, $c = a - Ad$

5.3 Incremental Dynamical Movement Primitive

5.3.1 Basic Incremental Dynamical Movement Primitive

Similar to (5.1), we define a new skill expressed by DMP in (5.10) that is different from the skills learned from the original demonstration. Given new position x^N and velocity v^N, the new skill is expressed as

$$\begin{cases} \tau\dot{v}^N = K\left(g - x^N\right) - Dv^N + (g - x_0) f^N(s) \\ \tau\dot{x}^N = v^N \end{cases}, \tag{5.10}$$

where $f^N(s) = \left(\theta^N\right)^T \Psi^N(s)$ is a new nonlinear function to be learned, and τ, K and D are as the same as those in (5.1).

For a new trajectory, the previous method will redefine a new pair of vectors θ^N and $\Psi^N(s)$ or add new kernels to DMP to adapt to novel situations [24]. DMP+ smartly reused and modified the kernels and weights for skill efficient adaptation to avoid re-calculation [23, 24]. In IDMP, we create extended feature terms $\eta_j(s)$ by making a linear transformation to the feature vector $\Phi(s) = [\varphi_1(s), \varphi_2(s), \ldots, \varphi_n(s)]$

$$\eta_j(s) \equiv \zeta_j\left(\Phi(s)W_{ej} + \beta_{ej}\right), j = 1, 2, \ldots, m \tag{5.11}$$

where ζ_j represents the jth transformation function of $\Phi(s)$, and W_{ej} and β_{ej} are random weights and bias terms. Then the new feature terms $\eta_j(s)$ are used in combination with $\hat{\psi}$ to achieve $\hat{\psi}$ in (5.12) contribute to $f^N(s)$ (see in Fig. 5.2).

It is obvious that $\sum_{j=1}^{m} \eta_j(s) + \sum_{i=1}^{n} \psi_i \neq s$, such that the property (5.8) is not satisfied, but the property (5.8) is the insurance that the final state value x converging

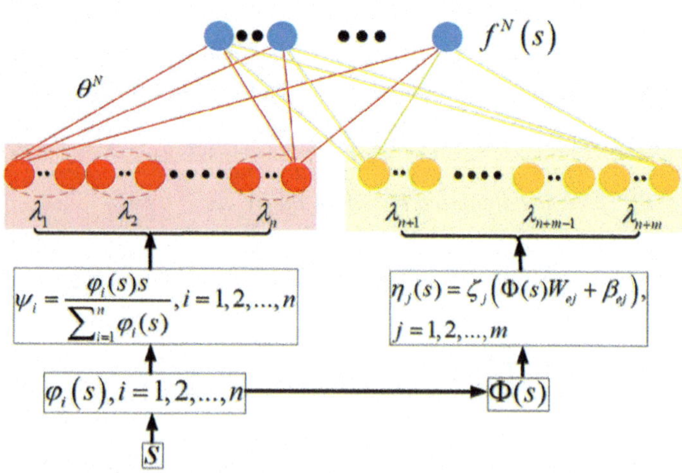

Fig. 5.2 Diagram of incremental dynamical movement primitive

monotonically to g. Therefore, using the new term $\eta_j(s)$ and ψ_i, we build a new term λ_k as

$$\lambda_k = \frac{q_k(s)\varphi_i(s) + (1 - q_k(s))\eta_j(s)}{\sum_{i=1}^{n}\varphi_i(s) + \sum_{j=1}^{m}\eta_j(s)}s, \quad k = 1, \ldots, m+n \tag{5.12}$$

where $q_k(s) = 1$,if $k \in [1, n]$ and $q_k(s) = 0$, if $k \in [n+1, n+m]$. It is obvious $\sum_{k=1}^{m+n}\lambda_k = s$, and if $k \in [n+1, n+m]$, (5.12) can be simplified as

$$\lambda_k = \frac{\eta_j(s)}{\sum_{i=1}^{n}\varphi_i(s) + \sum_{j=1}^{m}\eta_j(s)}s, \tag{5.13}$$

which is a modified term of $\eta_j(s)$ and equals to the Φ_j in (5.8). If $k \in [1, n]$, (5.12) can be simplified as

$$\begin{aligned}\lambda_k &= \frac{\varphi_i(s)}{\sum_{i=1}^{n}\varphi_i(s) + \sum_{j=1}^{m}\eta_j(s)}s \\ &= \frac{\varphi_i(s)s}{\sum_{i=1}^{n}\varphi_i(s)} \cdot \frac{\sum_{i=1}^{n}\varphi_i(s)}{\sum_{i=1}^{n}\varphi_i(s) + \sum_{j=1}^{m}\eta_j(s)} \\ &= \frac{\sum_{i=1}^{n}\varphi_i(s)}{\sum_{i=1}^{n}\varphi_i(s) + \sum_{j=1}^{m}\eta_j(s)}\psi_k,\end{aligned} \tag{5.14}$$

which means the term ψ_k is scaled up $\frac{\sum_{i=1}^{n}\varphi_i(s)}{\sum_{i=1}^{n}\varphi_i(s)+\sum_{j=1}^{m}\eta_j(s)}$ times and can be seen as the $\hat{\psi}_k$ in (5.8) to represent the modified ψ_k. Furthermore, we set a new scaling variable Γ as

$$\Gamma = \frac{\sum_{i=1}^{n}\varphi_i(s) + \sum_{j=1}^{m}\eta_j(s)}{\sum_{i=1}^{n}\varphi_i(s)}. \tag{5.15}$$

Then, we can get $\sum_{i=1}^{n}\varphi_i(s) + \sum_{j=1}^{m}\eta_j(s) = \frac{\Gamma}{\Gamma-1}\sum_{j=1}^{m}\eta_j(s)$, and λ_k in (5.12) is rewritten as

$$\begin{aligned}\lambda_k(s) &= \frac{q_k(s)\varphi_i(s)s}{\sum_{i=1}^{n}\varphi_i(s) + \sum_{j=1}^{m}\eta_j(s)} + \frac{(1 - q_k(s))\eta_j(s)s}{\sum_{i=1}^{n}\varphi_i(s) + \sum_{j=1}^{m}\eta_j(s)} \\ &= \frac{q_k(s)\varphi_i(s)s}{\Gamma\sum_{i=1}^{n}\varphi_i(s)} + \frac{(\Gamma - 1)(1 - q_k(s))\eta_j(s)s}{\Gamma\sum_{j=1}^{m}\eta_j(s)} \\ &= \frac{q_k(s)\psi_i(s)}{\Gamma} + \frac{(\Gamma - 1)(1 - q_k(s))\eta_j(s)s}{\Gamma\sum_{j=1}^{m}\eta_j(s)}.\end{aligned} \tag{5.16}$$

Here, we set

$$\gamma_j(s) = \frac{\eta_j(s)s}{\sum_{j=1}^{m} \eta_j(s)}, \tag{5.17}$$

Then λ_k in (5.16) is further expressed as

$$\begin{aligned}\lambda_k(s) = & \underbrace{q_k(s)\frac{\psi_i(s)}{\Gamma}}_{modified\ original\ skill} + \underbrace{(1-q_k(s))\frac{(\Gamma-1)\,\gamma_j(s)s}{\Gamma}}_{incremental\ skill} \\ = & \underbrace{q_k(s)\psi_i(s)}_{original\ skill} + \underbrace{q_k(s)\frac{(1-\Gamma)\,\psi_i(s)}{\Gamma}}_{old\ skill\ modification} + \underbrace{(1-q_k(s))\frac{(\Gamma-1)\,\gamma_j(s)s}{\Gamma}}_{incremental\ skill}.\end{aligned} \tag{5.18}$$

Remark 5.2 Different from trajectory modification methods by adding extra terms, e.g. $\gamma RV\varphi \exp(-\beta\varphi)$ in [15] ([V, φ are physical variables to represent velocity and joint), to the DMP function, λ_k and $f^N(s)$ are updated by adding new feature terms $\eta_j(s)$, which have nonlinear relationship with ψ_i in (5.2). The process is somewhat similar to the truncating kernels in DMP+. While the difference is, after adding $\eta_j(s)$, $f^N(s)$ can keep updating and the learned skills can approach the desired trajectory with any accuracy, which can be explained by the principles of incremental learning in [36]. However, DMP+ depends on but is constrained by the limited kernels. The compare of two methods will be further performed through the following experiment.

Seen from (5.18), if $m = 0$, we have $\lambda_i(s) = \psi_i(s)$, which is a standard DMP term. After adding more new terms of $\gamma_j(s)$, the effect of $\psi_i(s)$ changes and the number of $\lambda_k(s)$ increases. Here, we set θ^N and $\Psi^N(s)$ in (5.10) as $\theta^N = [w_1, w_2, \ldots, w_n, w_{n+1}, \ldots, w_{n+m}]^T$ and $\Psi(s) = [\lambda_1, \lambda_2, \ldots, \lambda_{m+n}]^T$. Each λ_k consists of three parts that are marked as the original skill [$q_k(s)\psi_i(s)$, the old skill modification $q_k(s)\psi_i(s)$ with a coefficient $(1-\Gamma)/\Gamma$ and incremental skill generated by the normalized Gaussian function $\gamma_j(s)$. Setting $\Upsilon(s) = [\gamma_1(s), \ldots, \gamma_m(s)]$, then the new forcing function $f^N(s)$:

$$\begin{aligned}f^N(s) &= \sum_{k=1}^{m+n} (w_k)^T \lambda_k(s) \\ &= (\theta^N)^T \Psi^N(s) \\ &= \underbrace{\theta^T \Psi(s)}_{original\ skill} + \underbrace{(1/\Gamma - 1)(\theta^C)^T \Psi(s)}_{old\ skill\ modification} + \underbrace{(1 - 1/\Gamma)(\theta^U)^T \Upsilon(s)}_{enhancement\ skill},\end{aligned} \tag{5.19}$$

where θ^N and $\Psi^N(s)$ are new weight and state vectors, which are expressed as

$$\begin{aligned}\theta^N &= [w_1, \ldots, w_n | w_1^c, \ldots, w_n^c | u_1, \ldots, u_m]^T \\ &= [\theta | \theta^C | \theta^U]^T\end{aligned}, \tag{5.20}$$

$$\begin{aligned}\Psi^N(s) &= [\psi_1(s), \ldots, \psi_n(s)\,|(1/\Gamma - 1)\,\psi_1(s), \ldots, (1/\Gamma - 1)\,\psi_n(s) \\ &\quad |(1 - 1/\Gamma)\,\gamma_1(s), \ldots, (1 - 1/\Gamma)\,\gamma_m(s)]^T \\ &= [\Psi(s)\,|(1/\Gamma - 1)\,\Psi(s)\,|(1 - 1/\Gamma)\,\Upsilon(s)]^T,\end{aligned} \tag{5.21}$$

where $\theta^C = \left[w_1^c, w_2^c, \ldots, w_n^c\right]$ is the weight of $(1/\Gamma - 1)\,\Psi(s)$ and $\theta^U = [u_1, u_2, \ldots, u_m]$ is the weight of $(1 - 1/\Gamma)\,\Upsilon(s)$. Moreover, $f^N(s)$ in (5.19) can be expressed as

$$\begin{aligned}f^N(s) &= \theta^T\Psi(s) + (1/\Gamma - 1)\left(\left(\theta^C\right)^T\Psi(s) - \left(\theta^U\right)^T\Upsilon(s)\right) \\ &= f(s) + (1/\Gamma - 1)\left(\left(\theta^C\right)^T\Psi(s) - \left(\theta^U\right)^T\Upsilon(s)\right).\end{aligned} \tag{5.22}$$

The desired value of $(1/\Gamma - 1)\left(\left(\theta^C\right)^T\Psi(s) - \left(\theta^U\right)^T\Upsilon(s)\right)$ is $\Delta f^N(s) = f^{N\_Tar} - f(s)$, where $f(s)$ is calculated based on (5.1), and f^{N_Tar} is calculated based on the new trajectory x^N as

$$f^{N\_Tar} = \left(\tau\dot{v}^N - K\left(g - x^N\right) - Dv^N\right)/\left(g - x_0\right). \tag{5.23}$$

From (5.20) and (5.21), the old vectors θ and $\Psi(s)$ are preserved in θ^N and $\Psi^N(s)$ and easy to be recovered by reducing new added terms. On the other hand, the skills can keep updating by extending θ^N and $\Psi^N(s)$ by the new features. (5.22) shows that the new forcing function $f^N(s)$ is calculated based on the old $f(s)$, and we can use f^{N_Tar} and $f(s)$ to compute the desired value of the skill modification $\Delta f^N(s)$ and further to obtain θ^C and θ^U by pseudo-inverse calculations.

Setting the initial value of θ^C as $\Delta f^N(s)\,(\Psi(s))^+$, the weights θ^C and θ^U are updated by Lemma 5.1 as

$$\begin{cases}\theta^C = \theta^C - \frac{db^T\Gamma}{1-\Gamma}\Delta f^N(s) \\ \theta^U = \frac{b^T\Gamma}{1-\Gamma}\Delta f^N(s) \\ d = -(\Psi(s))^+\Upsilon(s) \\ c = (\Psi(s))^+\Psi(s)\Upsilon(s) - \Upsilon(s) \\ b^T = \begin{cases} c^+ & if\ \ c \neq 0 \\ (1 + d^Td)^{-1}d^T(\Psi(s))^+ & if\ \ c = 0\end{cases}\end{cases}. \tag{5.24}$$

5.3.2 *Incremental Dynamical Movement Primitive for Multiple Stylistic Skill Generalization*

IDMP can be applied to generalize multiple stylistic skills when the initial learning skill is not accurate and there are several possible generalization solutions. Considering that all the possible skills are generated based on common initialized features, it is better to allow these skills to share the same features but with different weights.

We first assume that the position and velocity terms in the multiple demonstrations are x_i^N and v_i^N, $i = 1, 2, \ldots, m$ and the variables are x_i and v_i in the initial demonstration. By using (5.1), we can get an initial skill x_i and v_i as well as the common feature nodes $\Psi(s)$ and the forcing function $f(s)$ in the standard DMP. The next step is to compute the common extended terms $\eta_j(s)$ and the multiple sets of θ^C and θ^U for different trajectories simultaneously.

Here, we set θ_k^C and θ_k^U, $k = 1, 2, \ldots, m$ as the vectors of the $k$thtrajectory and set $\Theta^C = \{\theta_1^C, \theta_2^C, \ldots, \theta_m^C\}$ and $\{\theta_1^U, \theta_2^U, \ldots, \theta_m^U\}$ as vectors of the combination of weights. Similar to (5.23), we set the new learned skill expressed by DMP as

$$\begin{cases} \tau \dot{v}_i^N = K\left(g - x_i^N\right) - D v_i^N + \left(g - x_0\right) f_i^N(s) \\ \tau \dot{x}_i^N = v_i^N \end{cases} . \tag{5.25}$$

The extended term $\eta_j(s)$ and new variable λ_k are computed in the same way as in (5.11) and (5.18) to achieve $\Upsilon^c(s) = [\gamma_1(s), \gamma_2(s), \ldots, \gamma_m(s)]$ and $\gamma_i(s) = \eta_i(s)s/\sum_{j=1}^m \eta_j(s)$. Then the ith target value of $f_i^{N\_Tar}(s)$ is

$$f_i^{N\_Tar} = \left(\tau \dot{v}_i^N - K\left(g - x_i^N\right) - D v_i^N\right) / \left(g - x_0\right), \tag{5.26}$$

and the $f_i^N(s)$ in (5.25) has a similar expression to $f^N(s)$ in (5.19) and (5.22) as

$$\begin{aligned} f_i^N(s) &= \left(\theta_i^N\right)^T \Psi(s) \\ &= \theta^T \Psi(s) + (1/\Gamma^c - 1)\left(\theta_i^C\right)^T \Psi(s) + (1 - 1/\Gamma^c)\left(\theta_i^U\right)^T \Upsilon^c(s) \\ &= f(s) + (1/\Gamma^c - 1)\left(\left(\theta_i^C\right)^T \Psi(s) - \left(\theta_i^U\right)^T \Upsilon^c(s)\right), \end{aligned} \tag{5.27}$$

where Γ^c is defined as same as Γ in (5.15). The desired value of the term $(1/\Gamma^c - 1)\left(\left(\theta_i^C\right)^T \Psi(s) - \left(\theta_i^C\right)^T \Upsilon^c(s)\right)$ in (5.27) is

$$\Delta f_i^N(s) = f_i^{N\_Tar}(s) - f(s), \tag{5.28}$$

where $\Upsilon^c(s)$ represents the common extended feature vector, then the term θ_i^j, $i = 1, 2, \ldots, m$, $j = C, U$ in a vector and Θ^j, $j = C, U$ are calculated based on the common terms $\Upsilon^c(s)$ and Γ^c, similar to $\Upsilon(s)$ and Γ in (5.19), as

$$\begin{cases} \theta_i^C = \theta_i^C - \frac{d b^T \Gamma^c}{1 - \Gamma^c} \Delta f_i^N(s) \\ \theta_i^U = \frac{b^T \Gamma^c}{1 - \Gamma^c} \Delta f_i^N(s) \\ d = -(\Psi(s))^+ \Upsilon^c(s) \\ c = (\Psi(s))^+ \Psi(s) \Upsilon^c(s) - \Upsilon^c(s) \\ b^T = \begin{cases} c^+ & if \;\; c \neq 0 \\ (1 + d^T d)^{-1} d^T (\Psi(s))^+ & if \;\; c = 0 \end{cases} \end{cases} . \tag{5.29}$$

Using (5.29), we can get m group vectors of the weight set $[\theta_i^C, \theta_i^U]$ to express m stylistic skills. The detailed calculation procedure is realized by the pseudo code shown in Algorithm 5.1, for single and multiple stylistic incremental skill learning.

Algorithm 5.1 Incremental Dynamic Movement Primitive

Require: Parameters K, D, τ, number of kernels n, Gaussian function $\psi_i, i = 1, 2, \ldots, n$.
Transformation function $\xi_j(*)$, $j = 1, 2, \ldots, m$.
Start x_0 and end g of the trajectory.
Demonstrations:

- Single style: x, x^N
- Multi-style: x, x_i^N
- Convergence condition δ.

Ensure: Single style skill learning: Return the output weight θ^N (including θ, θ^C, and θ^U) and new skill in (5.10).
Multi-style skill learning: Return the output weight Θ^N (including θ, Θ^C, and Θ^U) and multiple learned skills in (5.25).

Step 1: Initial Skill Learning Using DMP
Initialize $s, h_i, c_i, i = 1, 2, \ldots, n$, and θ, ψ_i.
Calculate $f(s)$ using equation (5.4) and f_k^{Tar} in equation (5.5).
Finalize θ and $\varphi_i(s)$.
Step 2: Calculate Terms of the Extensive Skill Features
Set the feature mapping group $\Phi(s) = [\varphi_1(s), \ldots, \varphi_n(s)]$.
For $j = 1$:
repeat:
  Randomly select W_{ej}, β_{ej} and calculate $\eta_j(s) \equiv \zeta_j(\Phi(s)W_{ej} + \beta_{ej})$.
  Calculate Γ in equation (5.15) based on $\varphi_i(s)$ and $\eta_j(s)$.
  Calculate $\gamma_j(s)$ in equation (5.16) and the new feature term λ_k in equation (5.17).
until $j < m$
Step 3: New Single or Multiple Stylistic Skill Learning
Initialize the basic functions $\Psi(s)$ and Γ^c, set $m = 1$.
Calculate f^{N_Tar} in equation (5.23) and $\Delta f^N(s)$ in equation (5.22) for single style skill learning or $f_i^{N\_Tar}$ in equation (5.26) and $\Delta f_i^N(s)$ in equation (5.28) for multi-style skill learning.
repeat
  Randomly select $W_{e(j+1)}, \beta_{e(j+1)}$.
  Calculate $\eta_{j+1}(s) = \zeta_{j+1}(\Phi(s)W_{e(j+1)} + \beta_{e(j+1)})$ and add $\gamma_{j+1}(s)$ to extend $\Psi^N(s)$ in equation (5.21) and reinitialize θ^N in single style skill learning or Θ^U in multi-style skill learning.
  Use equations (5.24) and (5.29) to update θ^C, θ^U, and related terms.
  Use $f^N(s) = (\theta^N)^T\Psi^N(s)$ or $f_i^N(s) = (\theta_i^N)^T\Psi^N(s)$ to get $f^N(s)$ in single style skill learning or $f_i^N(s)$ in multi-style skill learning.
  Increment m.
until The error term $e = \sum_{i=1}^k |x - x_i^N|$ does not reach the threshold δ.

Remark 5.3 Since the IDMP-based multi-skill learning are based on the common features, the learned multiple skill can be transformed between each other by only changing the weight vectors. For example, we set the common state vector for two stylistic skills as $\Psi^N(s)=[\Psi(s)\,|(1/\Gamma - 1)\;\Psi(s)|\;\;(1 - 1/\Gamma)\;\Upsilon^c(s)]^T$ and weight vectors are $\theta_1^N = \left[\theta \left|\theta_1^C \right|\theta_1^U\right]^T$ and $\theta_2^N = \left[\theta \left|\theta_2^C \right|\theta_2^U\right]^T$ for two skills with the same length separately. The transformation of the two skills can be realized by linear interpolation [37] as

$$\begin{cases} \tau\dot{v}^N = K\left(g - x^N\right) - Dv^N + \left(g - x_0^N\right) f^N(s) \\ \tau\dot{x}^N = v^N \\ f^N(s) = \left(\theta^N\right)^T \Psi^N(s) \\ \theta^N = \alpha\theta_1^N + (1-\alpha)\,\theta_2^N \end{cases}, \tag{5.30}$$

where $\alpha \in (0, 1)$ is an adaptive factor to enable θ^N to change from θ_1^N to θ_2^N.

5.4 Incremental Adaptive Neural Network Control

The trajectory is replanned using IDMP method in the section above. However, as the trajectory changes, the controller's setpoints and performance limits also need to be changed. In this section, we will propose an new incremental adaptive NN control method to accumulate and reuse the learned skill, considering the limitations of tracking errors and robot dynamics estimation errors.

5.4.1 *System Dynamics Model and Control Objectives*

The dynamic model of robot system is expressed in a Lagrange-Euler form as

$$M(q)\ddot{q} + C(q, \dot{q})\dot{q} + G(q) = \tau + \tau_e, \tag{5.31}$$

where $q \in R^n$ is the simplification of $q(t)$ at time $t \in R^+$ and represents the joint information of robot arm, $M(q) \in R^{n\times n}$ is the inertia matrix, $C(q, \dot{q}) \in R^{n\times n}$ is the Coriolis and centrifugal torque matrix, and $G \in R^n$ is the gravitational torque. The control torques is τ and τ_e is a torques calculated by $\tau_e = J_e^T(q)F_e$, and F_e is forces exerted by the environment, and $J_e^T(q)$ is a Jacobian matrix. Setting the position of the end effector is x, then the relationship of q and x is $x(t) = \phi(q(t))$, $\dot{x}(t) = J(q(t))\dot{q}(t)$,where $\phi(*)$ is a function for joint and position transformation and $J(q(t))$ is a Jacobian matrix for robot system. The desired value of x is set as x^d, which is achieved by using traditional DMP or I-DMP. Using x^d, we can calculate q^d and set the tracking error of q to q^d as $e = q^d - q$. The desired tracking performance is to enable e to keep within the predesigned performance $-k_1 v(t) \le e(t) \le k_2 v(t)$, where k_1 and k_2 are constants and $v(t)$ is usually set as an exponential decaying performance function.

The dynamics system satisfies the following properties and assumptions:

Property 5.1 *The matrices* $\dot{M}(q) - C(q, \dot{q})$ *in (5.31) is skew -symmetric.*

5.4.2 Incremental Adaptive Neural Network Control

In order to realize the predesigned performance, the system controller is designed as

$$\tau = M\left(q^d\right)\dot{\alpha} + C\left(q^d, \dot{q}^d\right)\alpha + G\left(q^d\right) + K_\tau\left(\alpha - \dot{q}\right) - \hat{\tau}_e - g\left(\bar{S}\left(z\right)\right), \quad (5.32)$$

where $\alpha = \dot{q}^d - Le$,and L is a factor calculated in the following equation, K_τ is a positive constant factor, and $\hat{\tau}_e = J_e^T(q)\hat{F}_e$ and $\hat{F}_e$ represents the estimations of τ_e and F_e separately. Usually, the estimation error term $\tilde{\tau}_e = \tau_e - \hat{\tau}_e$ is coupled with uncertainties and disturbances [28, 29] to achieve a complex term $Y\left(q, \dot{q}, e, \dot{e}, \tilde{\tau}_e\right) = \tilde{M}\dot{\alpha} + \tilde{C}\alpha + \tilde{G} + \tilde{\tau}_e$, where $\tilde{M} = M(q) - M\left(q^d\right)$, $\tilde{C} = C(q, \dot{q}) - C\left(q^d, \dot{q}^d\right)$ and $\tilde{G} = G(q) - G\left(q^d\right)$ are the uncertain terms caused by joint tracking errors, $\tilde{\tau}_e$ represents the estimation error of the contact torque. $g(\bar{S}(z))$ is an incremental neural networks term with an expression of $g(\bar{S}(z)) = \hat{W}^T\bar{S}(z)$, where $\hat{W}$ is an estimated weight vector and $\bar{S}(z)$ represents a vector consisted of multiple Gaussian functions. We use $g(\bar{S}(z))$ to approach the error term $Y\left(q, \dot{q}, e, \dot{e}, \tilde{\tau}_e\right) = W^{*T}\bar{S}(z) + \varepsilon(z)$ that is expressed by the compositions of the desired weight vector W^{*T} and $\bar{S}(z)$, where $\varepsilon(z)$ is the approximation error of the neural network with the limitation of $\|\varepsilon(z)\| \leq \varepsilon*$, $\varepsilon* > 0$.

Taking (5.32) into (5.31), we have

$$\begin{aligned} M(q^d)\dot{\alpha} - M(q)\ddot{q} = &-C(q^d, \dot{q}^d)\alpha + C(q, \dot{q})\dot{q} \\ &- K_3(\alpha - \dot{q}) - \tau_e + \hat{\tau}_e \\ &- G(q^d) - G(q) + g(\bar{S}(z)), \end{aligned} \quad (5.33)$$

According to definition of α, we have $\dot{\alpha} = \ddot{q}^d - \dot{L}e - L\dot{e}$ and take it into (5.33), then

$$\begin{aligned} M(q)(\dot{\alpha} - \ddot{q}) &= -C(q, \dot{q})\alpha + C(q, \dot{q})\dot{q} + K_\tau(\alpha - \dot{q}) + \tilde{M}\dot{\alpha} + \tilde{C}\alpha + \tau_e - \hat{\tau}_e \\ &\quad + G(q^d) - G(q) - g(\bar{S}(z)), \\ &= -C(q, \dot{q})(\alpha - \dot{q}) + K_\tau(\alpha - \dot{q}) + \tilde{M}\dot{\alpha} + \tilde{C}\alpha + \tilde{G} + \tilde{\tau}_e - g(\bar{S}(z)), \\ &= -C(q, \dot{q})(\alpha - \dot{q}) + K_\tau(\alpha - \dot{q}) + Y(q, \dot{q}, e, \dot{e}, \tilde{\tau}_e) - g(\bar{S}(z)). \end{aligned} \quad (5.34)$$

To realize the predesigned performance and ensure the system stability, we set $\delta = \alpha - \dot{q}$ as the velocity-level tracking error to the virtual control term α and build the following barrier Lyaponov function as

$$
\begin{aligned}
V_q &= V_1 + V_2 \\
&= \sum_{i=1}^{n} h_i \ln\left(\frac{(k_2 v(t))^2}{(k_2 v(t))^2 - e^2}\right) \\
&\quad + \sum_{i=1}^{n} (1 - h_i) \ln\left(\frac{(k_1 v(t))^2}{e^2 - (k_1 v(t))^2}\right) \\
&\quad + \frac{1}{2} \delta^T M(q) \delta,
\end{aligned} \tag{5.35}
$$

where $V_2 = \frac{1}{2}\delta^T M(q)\delta$ and V_1 are the items in the rest of (5.35), and h_i is defined as

$$
h_i = \begin{cases} 1 & e > 0 \\ 0 & e \leq 0 \end{cases}. \tag{5.36}
$$

It is obvious that $V_q > 0$ and the time derivative of V_q is expressed as

$$
\begin{aligned}
\dot{V}_q &= \frac{1}{2}\sum_{i=1}^{n} h_i A_i + \frac{1}{2}\sum_{i=1}^{n}(1 - h_i) B_i + \dot{\delta}^T M(q)\delta + \frac{1}{2}\delta^T \dot{M}(q)\delta, \\
A_i &= \frac{(k_2 v(t))^2 - e^2}{(k_2 v(t))^2} \cdot \frac{2k_2 v(t)\dot{v}(t)((k_2 v(t))^2 - e^2) - (k_2 v(t))^2(2k_2 v(t)\dot{v}(t) - 2e\dot{e})}{((k_2 v(t))^2 - e^2)^2}, \\
B_i &= \frac{e^2 - (k_1 v(t))^2}{(k_1 v(t))^2} \cdot \frac{2k_1 v(t)\dot{v}(t)(e^2 - (k_1 v(t))^2) - (k_1 v(t))^2(2e\dot{e} - 2k_1 v(t)\dot{v}(t))}{(e^2 - (k_1 v(t))^2)^2}, \\
&= \sum_{i=1}^{n} h_i \left(\frac{\dot{v}(t)((k_2 v(t))^2 - e^2) - k_2 v(t)(k_2 v(t)\dot{v}(t) - e\dot{e})}{k_2 v(t)((k_2 v(t))^2 - e^2)}\right) \\
&\quad + \sum_{i=1}^{n}(1 - h_i)\left(\frac{\dot{v}(t)(e^2 - (k_1 v(t))^2) - k_1 v(t)(e\dot{e} - k_1 v(t)\dot{v}(t))}{k_1 v(t)(e^2 - (k_1 v(t))^2)}\right) \\
&\quad + \dot{\delta}^T M(q)\delta + \frac{1}{2}\delta^T \dot{M}(q)\delta \\
&= \sum_{i=1}^{n} h_i \frac{k_2 e\dot{e} v(t) - e^2 \dot{v}(t)}{v(t)((k_2 v(t))^2 - e^2)} \\
&\quad + \sum_{i=1}^{n}(1 - h_i)\frac{e^2 \dot{v}(t) - k_1 e\dot{e} v(t)}{v(t)(e^2 - (k_1 v(t))^2)} \\
&\quad + \dot{\delta}^T M(q)\delta + \frac{1}{2}\delta^T \dot{M}(q)\delta.
\end{aligned} \tag{5.37}
$$

According to the definition of α, we have $\dot{q}^d{=}\alpha{+}Le{=}\dot{e}{+}\dot{q}$, then $\dot{e}{=}\alpha{+}Le-\dot{q}$, then

$$
\begin{aligned}
\dot{V}_q &= \sum_{i=1}^{n} h_i \frac{k_2 e(\alpha + Le - \dot{q})v(t) - e^2\dot{v}(t)}{v(t)((k_2 v(t))^2 - e^2)} \\
&\quad + \sum_{i=1}^{n}(1-h_i)\frac{e^2\dot{v}(t) - k_1 e(\alpha + Le - \dot{q})v(t)}{v(t)(e^2 - (k_1 v(t))^2)} \\
&\quad + \dot{\delta}^T M(q)\delta + \frac{1}{2}\delta^T \dot{M}(q)\delta \\
&= \sum_{i=1}^{n} h_i \frac{k_2 e(\alpha - \dot{q})v(t) + k_2 L e^2 v(t) - e^2\dot{v}(t)}{v(t)((k_2 v(t))^2 - e^2)} \\
&\quad + \sum_{i=1}^{n}(1-h_i)\frac{e^2\dot{v}(t) - k_1 e(\alpha - \dot{q})v(t) + L k_1 e^2 v(t)}{v(t)(e^2 - (k_1 v(t))^2)} \\
&\quad + \dot{\delta}^T M(q)\delta + \frac{1}{2}\delta^T \dot{M}(q)\delta \\
&= \sum_{i=1}^{n} h_i \frac{k_2 e(\alpha - \dot{q})}{((k_2 v(t))^2 - e^2)} \\
&\quad + \sum_{i=1}^{n} h_i \frac{k_2 L e^2 v(t) - e^2\dot{v}(t)}{v(t)((k_2 v(t))^2 - e^2)} \\
&\quad + \sum_{i=1}^{n}(1-h_i)\frac{-k_1 e(\alpha - \dot{q})}{(e^2 - (k_1 v(t))^2)} \\
&\quad + \sum_{i=1}^{n}(1-h_i)\frac{e^2\dot{v}(t) + L k_1 e^2 v(t)}{v(t)(e^2 - (k_1 v(t))^2)} \\
&\quad + \dot{\delta}^T M(q)\delta + \frac{1}{2}\delta^T \dot{M}(q)\delta \\
&= \sum_{i=1}^{n}\left[\frac{h_i k_2 e}{((k_2 v(t))^2 - e^2)} - \frac{k_1(1-h_i)e}{(e^2 - (k_1 v(t))^2)}\right](\alpha - \dot{q}) \\
&\quad + \sum_{i=1}^{n} h_i\left(k_2 L - \frac{\dot{v}(t)}{v(t)}\right)\frac{e^2}{((k_2 v(t))^2 - e^2)} \\
&\quad + \sum_{i=1}^{n}(1-h_i)\left(\frac{\dot{v}(t)}{v(t)} + L k_1\right)\frac{e^2}{(e^2 - (k_1 v(t))^2)} \\
&\quad + \dot{\delta}^T M(q)\delta + \frac{1}{2}\delta^T \dot{M}(q)\delta.
\end{aligned}
\tag{5.38}
$$

If we set as $L = -\sqrt{\left(\frac{1}{k_2}\right)^2 + \left(\frac{1}{k_1}\right)^2 + (k_c)^2} \left\| \frac{\dot{v}(t)}{v(t)} \right\|$, since $\frac{e^2}{(k_2 v(t))^2 - e^2} > 0$ and $\frac{e^2}{e^2 - (k_1 v(t))^2} > 0$, then we have

$$\begin{aligned}
&\sum_{i=1}^{n} h_i \left(\frac{k_2 L - \dot{v}(t)}{v(t)} \right) \frac{e^2}{(k_2 v(t))^2 - e^2} + \sum_{i=1}^{n} (1 - h_i) \left(\frac{\dot{v}(t)}{v(t)} + L k_1 \right) \frac{e^2}{e^2 - (k_1 v(t))^2} \\
&< -k_c \left(\sum_{i=1}^{n} h_i \ln \left(\frac{(k_2 v(t))^2}{(k_2 v(t))^2 - e^2} \right) + \sum_{i=1}^{n} (1 - h_i) \ln \left(\frac{(k_1 v(t))^2}{e^2 - (k_1 v(t))^2} \right) \right) \\
&= -k_c V_1
\end{aligned} \tag{5.39}$$

Following the inequality $-\frac{e^2}{(k_2 v(t))^2 - e^2} \leq -\ln\left(\frac{(k_2 v(t))^2}{(k_2 v(t))^2 - e^2}\right)$ and $-\frac{e^2}{e^2 - (k_1 v(t))^2} \leq -\ln\left(\frac{(k_1 v(t))^2}{e^2 - (k_1 v(t))^2}\right)$, (5.39) can be expressed as

$$\begin{aligned}
&\sum_{i=1}^{n} h_i \left(k_2 L - \frac{\dot{v}(t)}{v(t)} \right) \frac{e^2}{(k_2 v(t))^2 - e^2} + \sum_{i=1}^{n} (1 - h_i) \left(\frac{\dot{v}(t)}{v(t)} + L k_1 \right) \frac{e^2}{e^2 - (k_1 v(t))^2} \\
&< -k_c \left(\sum_{i=1}^{n} h_i \ln \left(\frac{(k_2 v(t))^2}{(k_2 v(t))^2 - e^2} \right) + \sum_{i=1}^{n} (1 - h_i) \ln \left(\frac{(k_1 v(t))^2}{e^2 - (k_1 v(t))^2} \right) \right) \\
&= -k_c V_1.
\end{aligned} \tag{5.40}$$

Then (5.38) can be simplified as

$$\begin{aligned}
\dot{V}_q \leq & -k_c V_1 + \sum_{i=1}^{n} \left[\frac{h_i k_2 e}{(k_2 v(t))^2 - e^2} - \frac{k_1 (1 - h_i) e}{e^2 - (k_1 v(t))^2} \right] \delta \\
& + \dot{\delta}^T M(q) \delta + \frac{1}{2} \delta^T \dot{M}(q) \delta.
\end{aligned} \tag{5.41}$$

According to (5.34), we have $M(q)\,\dot{\delta} = -C(q, \dot{q})\,\delta + K_\tau \delta + Y(q, \dot{q}, e, \dot{e}, \tilde{\tau}_e) - g\left(\bar{S}(z)\right)$. Following Property 5.1, (5.41) can be simplified as

$$
\begin{aligned}
\dot{V}_q \leq & \sum_{i=1}^{n}\left[\frac{h_i k_2 e}{(k_2 v(t))^2 - e^2} - \frac{k_1(1-h_i)e}{e^2 - (k_1 v(t))^2}\right]\delta - \delta^T C(q,\dot{q})\delta \\
& + \delta^T Y(q,\dot{q},e,\dot{e},\tilde{\tau}_e) - \delta^T g(\bar{S}(z)) + \frac{1}{2}\delta^T \dot{M}(q)\delta - K_\tau \delta^T\delta - k_c V_1 \\
= & \sum_{i=1}^{n}\left[\frac{h_i k_2 e}{(k_2 v(t))^2 - e^2} - \frac{k_1(1-h_i)e}{e^2 - (k_1 v(t))^2}\right]\delta + \delta^T Y(q,\dot{q},e,\dot{e},\tilde{\tau}_e) \\
& - \delta^T g(\bar{S}(z)) - K_\tau \delta^T\delta - k_c V_1 \\
= & \left[\sum_{i=1}^{n}\left(\frac{h_i k_2 e}{(k_2 v(t))^2 - e^2} - \frac{k_1(1-h_i)e}{e^2 - (k_1 v(t))^2}\right) + Y(q,\dot{q},e,\dot{e},\tilde{\tau}_e)\right. \\
& \left. - g(\bar{S}(z))\right]\delta - K_\tau \delta^T\delta - k_c V_1.
\end{aligned}
\tag{5.42}
$$

According to the definition of h_i, we have $\frac{h_i k_2 e}{(k_2 v(t))^2 - e^2} \geq 0$ and $-\frac{k_1(1-h_i)e}{e^2 - (k_1 v(t))^2} \geq 0$, then

$$
\begin{aligned}
\max\left(\frac{k_2\|e\|}{(k_2 v(t))^2 - e^2}, \frac{k_1\|e\|}{e^2 - (k_1 v(t))^2}\right) \geq & \frac{h_i k_2 e}{(k_2 v(t))^2 - e^2} - \frac{k_1(1-h_i)e}{e^2 - (k_1 v(t))^2} \\
\geq & \ 0,
\end{aligned}
\tag{5.43}
$$

then we define $F_i = \frac{h_i k_2 e}{(k_2 v(t))^2 - e^2} - \frac{k_1(1-h_i)e}{e^2 - (k_1 v(t))^2}$ as a positive but limited term.

Then the updating rate of the weight vector $\hat{W}$ is

$$
\dot{\hat{W}} = \left(\Gamma_s \bar{S}(z) + u\tanh\frac{u\delta}{\varpi}\right)\delta - K_s \Gamma_s \hat{W}, \tag{5.44}
$$

where $\Gamma_s > \left\|\tilde{W}\right\|$ is a large positive matrix, tanh $(*)$ is a hyperbolic tangent function and K_s is positive factor.

We further create a quadratic term $V_m = \tilde{W}^T \Gamma_s^{-1} \tilde{W} > 0$ and the time derivative of V_m is

$$
\begin{aligned}
\dot{V}_m &= \tilde{W}^T \Gamma_s^{-1} \dot{\tilde{W}} \\
&= \tilde{W}^T \Gamma_s^{-1}\left(\left(-\Gamma_s \bar{S}(z) - u_i \tanh\tfrac{u_i\delta}{\varpi_i}\right)\delta + K_s \Gamma_s \hat{W}\right) \\
&= \tilde{W}^T \Gamma_s^{-1}\left(\left(-\Gamma_s \bar{S}(z) - u_i \tanh\tfrac{u_i\delta}{\varpi_i}\right)\delta + K_s \Gamma_s \left(W^* - \tilde{W}\right)\right) \\
&= -\tilde{W}^T \bar{S}(z)\delta - \tilde{W}^T \Gamma_s^{-1} u_i \tanh\tfrac{u_i\delta}{\varpi_i}\delta + \tilde{W}^T K_s \left(W^* - \tilde{W}\right).
\end{aligned}
\tag{5.45}
$$

Therefore for the hybrid Lyapunov function $V = V_q + V_m > 0$, the time derivative is expressed as

$$
\begin{aligned}
\dot{V} \leq & \left[\sum_{i=1}^{n} F_i + Y\left(q, \dot{q}, e, \dot{e}, \tilde{\tau}_e\right) - g\left(\bar{S}(z)\right)\right]\delta - \tilde{W}^T\bar{S}(z)\delta - \tilde{W}^T{\Gamma_s}^{-1}u_i\tanh\tfrac{u_i\delta}{\varpi_i}\delta + \\
& \tilde{W}^T K_s\left(W^* - \tilde{W}\right) - K_\tau\delta^T\delta - k_c V_1 \\
= & \left[\sum_{i=1}^{n} F_i - \tilde{W}^T{\Gamma_s}^{-1}u_i\tanh\tfrac{u_i\delta}{\varpi_i} + \varepsilon(z)\right]\delta + \tilde{W}^T K_s\left(W^* - \tilde{W}\right) - K_\tau\delta^T\delta - k_c V_1.
\end{aligned}
\tag{5.46}
$$

Following the Young's inequality and definition of Γ_s, we can obtain the following inequality [35]

$$
F\delta - \tilde{W}^T{\Gamma_s}^{-1}u\delta\tanh\frac{u\delta}{\varpi} \leq \iota\varpi. \tag{5.47}
$$

Thus (5.46) can be further deduced by expressing the weights by the extended matrices as $W^* = \left[W_{ori}^* \;\; W_{ex}^*\right]^T$ and $\tilde{W} = \left[\tilde{W}_{ori} \;\; \tilde{W}_{ex}\right]^T$, where W_{ori}^* and $\tilde{W}_{ori}$ represent the vectors of the original weight and weight error, and W_{ex}^* and $\tilde{W}_{ex}$ are the vectors of the extend features. Then

$$
\begin{aligned}
\dot{V} &\leq -k_c V_1 + \tfrac{1}{2}\varepsilon(z)^2 + \iota\varpi - K_\tau\delta^T\delta - \tfrac{1}{2}\tilde{W}^T K_s\tilde{W} + \tfrac{1}{2}W^{*T}K_s W^* \\
&= -k_c V_1 - K_\tau\delta^T\delta - \tfrac{1}{2}K_s\left\|\begin{matrix}\tilde{W}_{ori}\\ \tilde{W}_{ex}\end{matrix}\right\|^2 + \tfrac{1}{2}\varepsilon(z)^2 + \iota\varpi + \tfrac{1}{2}K_s\left\|\begin{matrix}W_{ori}^*\\ W_{ex}^*.\end{matrix}\right\|^2
\end{aligned}
\tag{5.48}
$$

Considering the completed expression of V is $V = V_1 + V_2 + V_m = V_1 + \delta^T M(q)\delta + \tilde{W}^T{\Gamma_s}^{-1}\tilde{W}$, then (5.48) can be expressed as

$$
\dot{V} = -\eta V + \sigma, \tag{5.49}
$$

where $\eta = \min\left(\lambda_{\min}(k_c), \frac{2\lambda_{\min}(K_\tau)}{\lambda_{\max}(M(q))}, \frac{2\lambda_{\min}(K_s)}{\lambda_{\max}\left({\Gamma_s}^{-1}\right)}\right)$ and $\sigma = \frac{1}{2}\varepsilon(z)^2 + \iota\varpi + \frac{1}{2}K_s \left\|\begin{matrix}W_{ori}^*\\ W_{ex}^*\end{matrix}\right\|^2$, and the solution is

$$
V(t) \leq \left(V(0) - \frac{\sigma}{\eta}\right)\exp(-\eta t) + \frac{\sigma}{\eta} \leq V(0)\exp(-\eta t) + \frac{\sigma}{\eta}. \tag{5.50}
$$

Since the terms $\varepsilon(z)$, $\iota\varpi$ and $K_s\left\|\left[W_{ori}^* \;\; \tilde{W}\right]^T\right\|^2$ are bounded and σ/η is bounded, then $V(t)$ is bounded and converged along with the time. This completes the proof.

Remark 5.4 In our previous research [27, 35], we proposed a combining framework of trajectory learning and board learning-based control to approximate the unknown dynamics of the robot. The improvements of this proposed method are building a new Lyapunov function and creating a new weight estimation function (5.44) based on the trajectory learned by IDMP. Therefore, the controller is designed with a specific parameters as L in the definition of α.

5.5 Experiments

We will verify the three contributions shown in the Introduction through three following experiments.

5.5.1 Experiment 1. Accurate Trajectory Approaching

In this experiment, we aim to verify the improvement of trajectory tracking accuracy by IDMP, compared with other DMP-based methods. We prepare a handwriting letter 'A' in blue in Fig. 5.3 and use the standard DMP, DMP+ proposed by Wang and Wu et al. in [24], and IDMP in this section to track this trajectory with the same kernels.

First we choose 5 kernels for the forcing function for all DMP-based methods. The 5 initial kernels are chosen with random centres and widths of the radial basis functions for all three methods. The initial learning results of DMP in Fig. 5.3a show a not good tracking effect.

With DMP+, the 5 kernels are modified and the mean squared error (MSE) in tracking the demonstrated trajectory is significantly reduced (see Fig. 5.3b). IDMP can extend the feature vector $\Psi(s)$ by adding new transformation terms $\eta_j(s)$ of the 5 original kernels in (5.11), so that the trajectory tracking performance is further improved (see Fig. 5.3c).

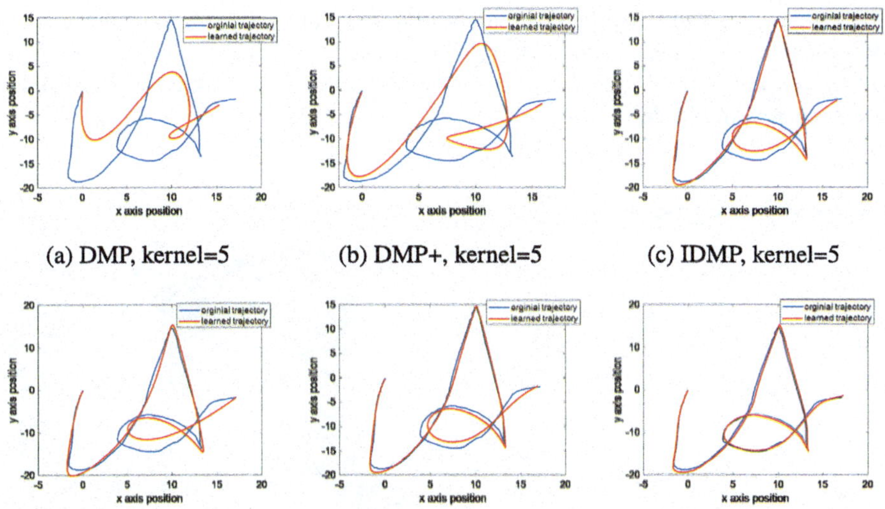

(a) DMP, kernel=5 (b) DMP+, kernel=5 (c) IDMP, kernel=5

(d) DMP, kernel=20 (e) DMP+, kernel=20 (f) DMP+, kernel=20

Fig. 5.3 Handwriting trajectory learning by using standard DMP, DMP+ and IDMP with different kernals

We further expand the number of initial kernels in three methods from 5 to 20 and the simulation results are shown in Fig. 5.3d–f. The tracking performances of all methods are significantly better than those with 5 kernels. The MSEs to the original trajectory of IDMP are much lower than the results of the previous two cases and the trajectory almost coincides with the demonstration, which benefits from the increasing number of the extended features and certifies that IDMP has the best trajectory tracking accuracy among the three methods. But, it also shows that the tracking errors are still partially affected by the initial number of kernels.

5.5.2 Experiment 2. Multi-style Skill Learning and Transformation

The second experiment is to examine multi-style skill learning, modification, and transformation. As shown in Fig. 5.4, we retrofitted a PHANTOM Desktop haptic device and fixed a pen at the end of the effector. The demonstrator operates the haptic device to write letters and the device records trajectories of the end tip. The trajectories are processed (e.g., alignment and filtering) and then used for handwriting style learning and transformation under the control of the incremental adaptive NN controller in (5.32).

As shown in Fig. 5.5a, we write five letters ‘a’, ‘z’, ‘l’, ‘k’ and ‘w’ with similar size and same start and end. The letter ‘a’ is selected as the initial writing style and the others are provided as the writing styles after skill modification and transformation. After matching and resampling the demonstrated trajectories, we use the method described in Algorithm 5.1 to learn stylistic letters and realize skill modification and transformation from ‘a’ to other letters. The processes for the skill transformations are presented in Fig. 5.6.

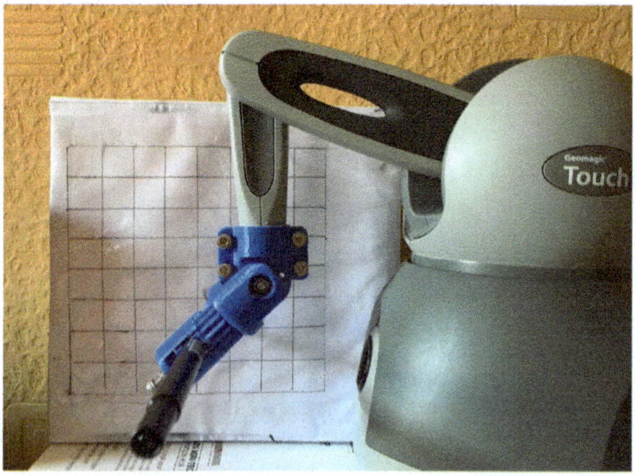

Fig. 5.4 Experimental setup

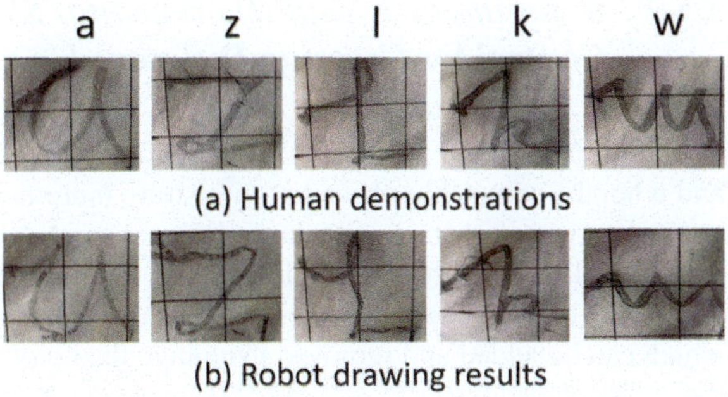

Fig. 5.5 Human demonstrations of handwriting and robot drawing results after multi-style skill learning

The initial trajectory 'a' is coloured red, marked with a red square in the centre of Fig. 5.6, and learned with the standard DMP. The targeted manuscripts are presented with black squares in the four corners. The transformation starts at centre 'a' to approach the handwritings in the corners, performing every 5 incremental steps. In this way, the shapes near the centre are more similar to 'a'. With the extension of feature vector and weight vector, the learned trajectories are gradually changed from 'a' to other stylistic letters. After adding 30 extended feature kernels, the modified letters are clearly distinguishable from each other, resulting in the letters in the corners of Fig. 5.6. For some letters, such as 'z' and 'k', the modified letters from 'a' have been similar to the final trajectories. While, others, such as 'w' and 'l', are changed gradually to the desired states. Since the common features are determined by all the stylistic letters, the trained trajectories that are closest to the corner demonstrations still have some differences that can be considered as compromise results. The skill transformation between different stylistic actions is realized by using (5.30) to change weight vectors. Finally, we utilize the haptic device as an actuator and use adaptive NN control method to draw the learning results, as shown in Fig. 5.5b.

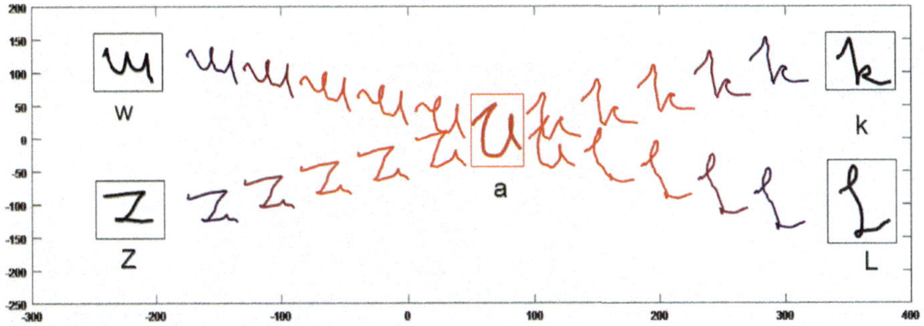

Fig. 5.6 Incremental learning process from the letter 'a' to multilple syles of handwritings:'w','z','k'and'l'

5.5.3 Experiment 3. Dual-Incremental Skill Learning and Control for Crossing Different-Height Obstacles

The third experiment is also conducted with the haptic device PHANTOM Desktop and a height-adjustable obstacle to illustrate incremental skill learning process and its application of obstacle avoidance in practice. As shown in Fig. 5.7, humans hold the joystick to cross the obstacle and put the pen tip to touch the intended target point. The current obstacle is consists of three 2 cm $\times$ 2 cm $\times$ 2 cm cubes, each of which can be added and removed to change the height of the obstacle. On the top of the cubes, we add a 2 cm $\times$ 8 cm $\times$ 0.2 cm lip. On the base plane, we set one start point and nine target points on the two sides of the obstacle. The central point on the right side is used for human demonstration and the other points, which are 2 cm away from each other, are used for skill generalization.

After collecting data from demonstrations (gray lines in Fig. 5.8a), we use DMP to learn an initial skill of crossing an obstacle with a height of 1 block (red line in Fig. 5.8a). In our previous work [26], the demonstrations for the height-adjustable obstacle are divided into several phases and the final positions of the internal phases are changed according to the heights of the new obstacles. In this section, we use the IDMP to realize skill modification. As shown in Fig. 5.8b, the deep green and dark blue lines are the final learning results for crossing two and three blocks. The lines with gradient colours from red to green and from green to blue show the learning process with the increase of extended features, according to the diagram in Algorithm 5.1. Figure 5.8b also verifies the convergence of the learning results such that the degrees of curve changes are decreasing until they approach the final learning results. After gaining the ability to overcome the obstacles of different heights, we can generalize them to achieve different goals by changing factor g in (5.25).

In Fig. 5.8c and d, the green lines show the learning effect for different obstacles and the red lines are the generalized trajectories of the robot end to achieve different goals.

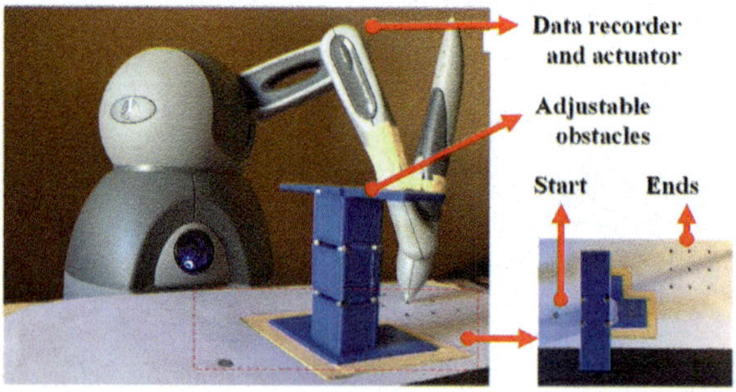

Fig. 5.7 Experimental setup for obstacle avoidance

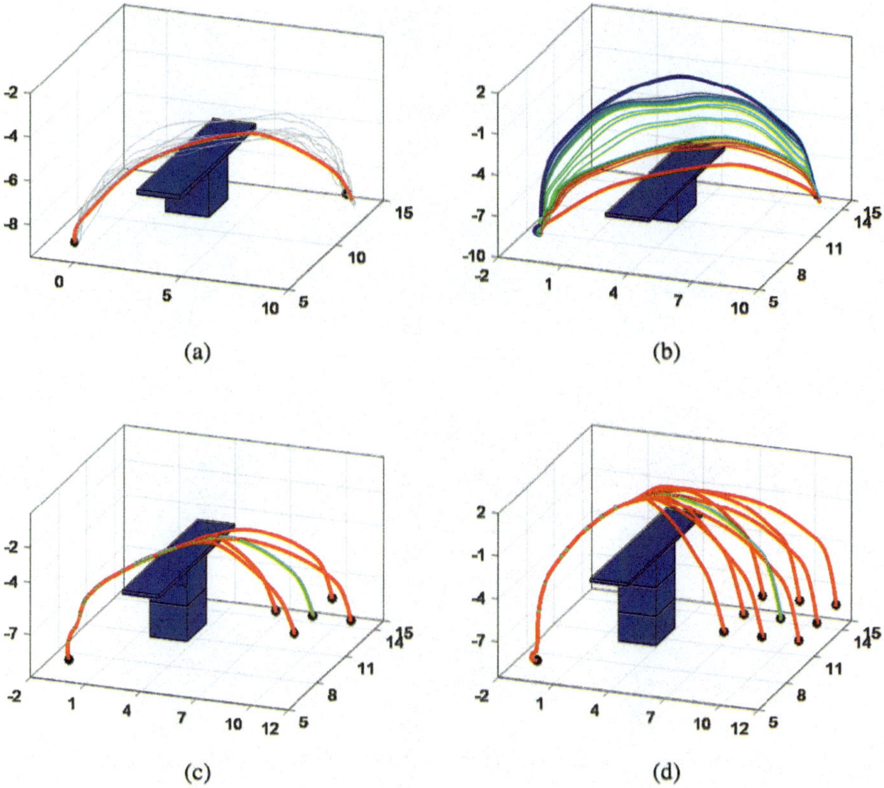

Fig. 5.8 Skill incremental learning and generalization based on IDMP and adaptive NN-based control method **a** Demonstrations and skill learning based on DMP (gray lines are trajectories of demonstrations and the red line is the learned skill) **b** Incremental learning process for different learning skills (red line is the crossing skill for 1 block height, green line is the crossing skill for 2-block height, and blue line is for the height of 3 blocks, and the thin lines between the different skills represent the transformation of the skills) **c** Generalization of the skill for the 2-block height (green line shows the skill learning and red lines represent the skill generalizations) **d** Generalization of the skill for the height of 3 blocks (green line shows the skill learning and red lines show the skill generalization)

Using the adaptive NN controller in (5.32), the joystick works as an actuator to follow the generalized skills with limited tracking errors. Figure 5.9a shows human demonstrations process. Figure 5.9b and c show the joystick movement to reach the predefined goals without conflicting with obstacles.

(a) Skill demonstrations

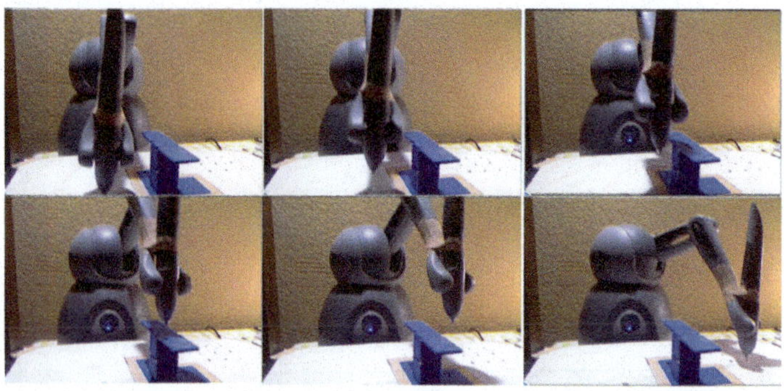

(b) Robot crosses 2-block height obstacle avoidance autonomously

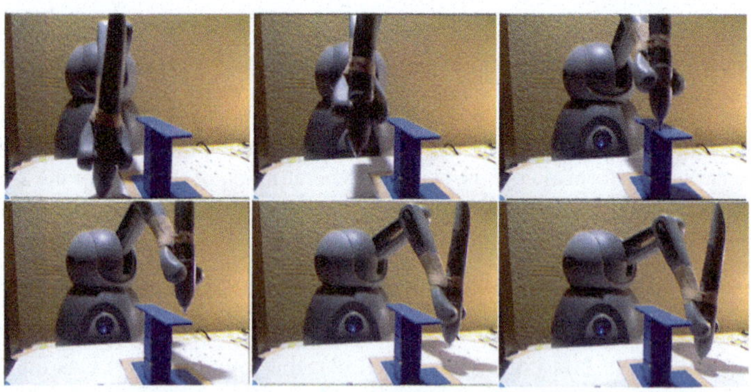

(c) Robot crosses 3-block height obstacle avoidance autonomously

Fig. 5.9 Skill generalization and control for height-adjustable obstacle avoidance based on human demonstrations

5.5.4 Discussion

The three experiments verify three properties of the incremental trajectory and force learning framework for robots proposed in the Introduction. Experiment 1 shows the advantage of IDMP in terms of accuracy, compared to DMP+ and standard DMP and can realize life-long skill learning to some extent. Experiment 2 shows the versatility of IDMP in generalizing and transforming skills into multiple styles. The incremental adaptive controller ensures system stability and keeps trajectory tracking errors within performance limits. Experiment 3 shows applications of the proposed framework in obstacle avoidance. Moreover, the properties can be used in combination to achieve other goals. For example, we can first generalize the original skill to multi-style skills, and then further refine the details of a specific style. Since the old features and weights are contained in the vectors, we can easily perform skill transformation from one to another or transformation between different skills as Remark 5.3 shown, and ensure the smoothness of the transformation by choosing an appropriate. But, since IDMP is calculated based on the elements of the original DMP skills, the proposed method is still limited by the original learning outcomes.

5.6 Conclusion

In this section, we propose a new framework for incremental trajectory and force learning and generalization to modify initially learned skills due to the environmental changes and inaccurate initialization. The trajectory learning part is based on the IDMP, and the controller uses adaptive NN control method to reduce the cost and improve the efficiency of learning. Three experiments are taken to verify the effectiveness and advantages of the proposed framework in accurate trajectory tracking, multi-stylistic trajectory tracking and application in obstacle avoidance. Compared to other DMP-based methods, this framework achieves better performance in robot trajectory tracking and greater flexibility in skill modification and transformation.

References

1. Argall BD, Chernova S, Veloso M, Browning B (2009) A survey of robot learning from demonstration. Rob Autonom syst 57(5):469–483
2. Ravichandar H, Polydoros AS, Chernova S, Billard A (2020) Recent advances in robot learning from demonstration. Ann Rev Control Rob Autonom Syst 3:297–330
3. Huang Y, Rozo L, Silvério J, Caldwell DG (2019) Kernelized movement primitives. Int J Rob Res 38(7):833–852
4. Deniša M, Petric T, Gams A, Ude A (2016) A review of compliant movement primitives. Rob Control 1–17
5. Saveriano M, Abu-Dakka FJ, Kramberger A, Peternel L (2021) Dynamic movement primitives in robotics: a tutorial survey. Int J Rob Res 42(13):1133–1184
6. Schaal S, Peters J, Nakanishi J, Ijspeert A (2003) Control, planning, learning, and imitation with dynamic movement primitives. Workshop on bilateral paradigms on humans and humanoids, IEEE international conference on intelligent robots and systems, Las Vegas, NV 2003:1–21
7. Ijspeert AJ, Nakanishi J, Hoffmann H, Pastor P, Schaal S (2013) Dynamical movement primitives: learning attractor models for motor behaviors. Neural Comput 25(2):328–373
8. Yuan Y, Li Z, Zhao T, Gan D (2019) DMP-based motion generation for a walking exoskeleton robot using reinforcement learning. IEEE Trans Indus Electron 67(5):3830–3839
9. Li Z, Zhao T, Chen F, Hu Y, Su CY, Fukuda T (2017) Reinforcement learning of manipulation and grasping using dynamical movement primitives for a humanoid like mobile manipulator. IEEE/ASME Trans Mechatron 23(1):121–131
10. Pervez A, Mao Y, Lee D (2017) Learning deep movement primitives using convolutional neural networks. Paper presented at the 2017 IEEE-RAS 17th international conference on humanoid robotics (Humanoids), Birmingham, UK, 15–17 November 2017
11. Yang C, Chen C, He W, Cui R, Li Z (2018) Robot learning system based on adaptive neural control and dynamic movement primitives. IEEE Trans Neural Netw Learning Syst 30(3):777–787
12. Kastritsi T, Dimeas F, Doulgeri Z (2018) Progressive automation with DMP synchronization and variable stiffness control. IEEE Rob Autom Lett 3(4):3789–3796
13. Yang C, Chen C, Wang N, Ju Z, Fu J, Wang M (2018) Biologically inspired motion modeling and neural control for robot learning from demonstrations. IEEE Trans Cogn Dev Syst 11(2):281–291
14. Abu-Dakka FJ, Kyrki V (2020) Geometry-aware dynamic movement primitives. Paper presented at the 2020 IEEE international conference on robotics and automation (ICRA), Paris, France, 31 May 2020–31 August 2020
15. Hoffmann H, Pastor P, Park DH, Schaal S (2009) Biologically-inspired dynamical systems for movement generation: automatic real-time goal adaptation and obstacle avoidance. Paper presented at the 2009 IEEE international conference on robotics and automation, Kobe, Japan, 12–17 May 2009
16. Han L, Kang P, Chen Y, Xu W, Li B (2019) Trajectory optimization and force control with modified dynamic movement primitives under curved surface constraints. Paper presented at the 2019 IEEE international conference on robotics and biomimetics (ROBIO), Dali, China, 06–08 December 2019
17. Lu Z, Wang N, Yang C (2022) A novel iterative identification based on the optimised topology for common state monitoring in wireless sensor networks. Int J Syst Sci 53(1):25–39
18. Gams A, Nemec B, Ijspeert AJ, Ude A (2014) Coupling movement primitives: Interaction with the environment and bimanual tasks. IEEE Trans Rob 30(4):816–830
19. Umlauft J, Sieber D, Hirche S (2014) Dynamic movement primitives for cooperative manipulation and synchronized motions. Paper presented at the 2014 IEEE international conference on robotics and automation (ICRA), Hong Kong, China, 31 May 2014–07 June 2014
20. Huang R, Cheng H, Qiu J, Zhang J (2019) Learning physical human-robot interaction with coupled cooperative primitives for a lower exoskeleton. IEEE Trans Autom Sci Eng 16(4):1566–1574

21. Saveriano M, Lee D (2018) Incremental skill learning of stable dynamical systems. Paper presented at the 2018 IEEE/RSJ international conference on intelligent robots and systems (IROS), Madrid, Spain, 01–05 October 2018
22. Lemme A, Reinhart RF, Steil JJ (2014) Self-supervised bootstrapping of a movement primitive library from complex trajectories. Paper presented at the 2014 IEEE-RAS international conference on humanoid robots, Madrid, Spain, 18–20 November 2014
23. Wu Y, Wang R, D'Haro LF, Banchs RE, Tee KP (2018) Multi-modal robot apprenticeship: Imitation learning using linearly decayed DMP+ in a human-robot dialogue system. Paper presented at the 2018 IEEE/RSJ international conference on intelligent robots and systems (IROS), Madrid, Spain, 01–05 October 2018
24. Wang R, Wu Y, Chan WL, Tee KP (2016) Dynamic movement primitives plus: for enhanced reproduction quality and efficient trajectory modification using truncated kernels and local biases. Paper presented at the 2016 IEEE/RSJ international conference on intelligent robots and systems (IROS), Daejeon, 09–14 October 2016
25. Schaal S (2006) Dynamic movement primitives-a framework for motor control in humans and humanoid robotics. In: Adaptive motion of animals and machines. Springer, Tokyo
26. Yang C, Zeng C, Cong Y, Wang N, Wang M (2018) A learning framework of adaptive manipulative skills from human to robot. IEEE Trans Indus Inf 15(2):1153–1161
27. Wang N, Chen C, Yang C (2020) A robot learning framework based on adaptive admittance control and generalizable motion modeling with neural network controller. Neurocomputing 390:260–267
28. Liu Z, Lin W, Yu X, Rodriguez-Andina JJ, Gao H (2021) Approximation-free robust synchronization control for dual linear motors driven systems with uncertainties and disturbances. IEEE Trans Indus Electron 69(10):10500–10509
29. Shi P, Sun W, Yang X, Rudas I, Gao H (2022) Master-slave synchronous control of dual drive gantry stage with cogging force compensation. IEEE Trans Syst Man Cybern Syst 53(1):216–225
30. Zhang Y, Li M, Yang C (2021) Robot learning system based on dynamic movement primitives and neural network. Neurocomputing 451:205–214
31. Si W, Wang N, Yang C (2023) Composite dynamic movement primitives based on neural networks for human-robot skill transfer. Neural Comput Appl 35(32):23283–23293
32. Akbulut M, Oztop E, Seker MY, Hh X, Tekden A and Ugur E (2021) Acnmp: skill transfer and task extrapolation through learning from demonstration and reinforcement learning via representation sharing. Paper presented at the proceedings of the 2020 conference on robot learning, PMLR, virtual, 16–18 November 2020
33. Liu Z et al (2017) Broad learning system: feature extraction based on K-means clustering algorithm. Paper presented at the 2017 4th international conference on information, cybernetics and computational social systems (ICCSS), Dalian, China, 24–26 July 2017
34. Matsubara T, Hyon SH, and Morimoto J (2010) Learning stylistic dynamic movement primitives from multiple demonstrations. Paper presented at the 2010 IEEE/RSJ international conference on intelligent robots and systems, Taipei, Taiwan, 18–22 October 2010
35. Huang H, Zhang T, Yang C, Chen CP (2019) Motor learning and generalization using broad learning adaptive neural control. IEEE Trans Indus Electron 67(10):8608–8617
36. Chen CLP, Liu Z (2017) Broad learning system: an effective and efficient incremental learning system without the need for deep architecture. IEEE Trans Neural Netw Learn Syst 29(1):10–24
37. Calinon S (2020) Gaussians on Riemannian manifolds: applications for robot learning and adaptive control. IEEE Rob Autom Mag 27(2):33–45

Part III
Bio-inspired Autonomous Learning for Dexterous Manipulations

Chapter 6
A Constrained DMP Framework for Robot Skills Learning and Generalization from Human Demonstrations

6.1 Introduction

Learning from demonstration (LfD), inspired by neuroscience, is an effective way for robotics to learn manipulation skills such as opening doors or grasping cups from human's natural actions [1]. Since the 1980s, plenty of methods have been proposed, e.g., the authors in [2–4] used the Gaussian mixture model (GMM) and Gaussian mixture regression (GMR) to regenerate motions. Ng and Russell [5] proposed the inverse reinforcement learning (IRL) method to build up an unknown reward function based on the observed trajectories to characterize solutions. Ijspeert et al. [6, 7] proposed the dynamic movement primitive (DMP) model that produces a trajectory similar to the demonstrations to solve the path planning without building kinematic models. Since then, the DMP model has been modified and improved widely for imitation and learning of human skills like reaching movements [8], object grasping [9] and clothes manipulation [10], etc.

Constraints in joint and task spaces [11] and caused by force interaction [12] are common topics for LfD. Some researchers improved the DMP model and combined neural networks control with DMP [13, 14] to solve the tasks with special constraints such as obstacle avoidance [15, 16, 38–41], bimanual operation [19, 20, 42–46], and interaction with external objects [17], the majority of which added a coupling term based on the basic functions.

For obstacle avoidance, Khansari-Zadeh et al. [38] and Park et al. [39] took repulsive potential fields as coupling terms into DMP for obstacle avoidance. Hoffmann et al. [16] motivated by biological data and human behaviors and modified the DMP model by adding an acceleration term to avoid moving obstacles. But, Pairet et al. [40] pointed out that the additional coupling term will bring some limitations: dead-zone compromising the method's reliability, without a strategy to guide to a preferred route to circumnavigate an obstacle, performance decrease facing non-point obstacles, time-consuming, and prone to measurement noise. Therefore, they proposed a hierarchical framework that combines the versatility of DMP and the

C. Yang et al., *Advanced Teleoperation and Robot Learning for Dexterous Manipulation*, Springer Tracts in Advanced Robotics 160,
https://doi.org/10.1007/978-3-031-78501-6_6

strengths of learning techniques [40]. As statements mentioned in [39], these proposed approaches enable a robot to avoid conflict with an obstacle by predefining policies. However, a nonexpert cannot teach a robot his/her special skills of avoiding a collision with different objects by demonstrations.

Cooperative DMP for bimanual robot manipulation [19, 20, 42–46] and multirobot formation [46] have another kind of constraint, which brings strong space constraints to all the related roles. Umlauft et al. [20] inspired by DMP interacting with external perturbations, added an error item to increase adaptability under external disturbances. Kulvicius et al. [44] studied interactive DMP and added an accelerating predictive reaction for the coupling agents, which is similar to the term for obstacle avoidance. Gams et al. [19] argued that force-based modulation of the DMP at both velocity and acceleration levels is needed for smoother interactions, which can be seen as the combination of above mentioned two papers. They also used the iterative learning control method to learn the coupling term to modify the original trajectory based on the force feedback generated during the executions of the tasks. Lee et al. [46] studied with the purpose of both obstacle avoidance and cooperative manipulation by DMP for aerial robot formation, and two terms were formulated based on regular DMP to solve two problems separately. Lastly, some researchers studied special constraints such as curved surface [47] to modify DMP by adding an accelerating coupling term for realizing trajectory planning and force control [17].

Seen from the presented work, it is not hard to find that the coupling terms are specially designed with different constraints and purposes for the varying tasks and operational demands. In fact, these terms have two implicit usages: path replanning and generalization. As there are many research works that have proposed path replanning methods for various scenes, we hope to propose a general DMP-based skill generalization method based on the path replanning methods to enable the original path to fit tasks with constraints, especially with several dispersed constraints. Some researchers have made a few experimental studies, e.g., in [35], Krug et al. combined the model predictive control (MPC) method and DMP to calculate possibilities of obstacle avoidance for online optimization of trajectory. Furthermore, Khansari-Zadeh and Billard [38] proposed an idea of a safety margin around obstacles via a safety factor for obstacle avoidance. Compared with other research works, these two papers attempted to provide a general DMP method for more cases with constraints. With this inspiration, we propose a novel DMP framework that combines the control concept of barrier Lyapunov functions (BLF) with a regular DMP model. The idea of integrating control and learning methods is similar to our previous work [18, 27, 28, 31, 32, 34, 48, 49] but has a difference that is by adding an acceleration term calculated by constrained paths and safety margins, the control method will be embedded in the DMP model and modify the trajectories timely. The main contributions are listed as follows.

1. Proposing a new abstracted DMP model with classified constraints based on the summary of plenty of improved DMP cases with constraints, covering obstacle avoidance, cooperative manipulation, and curved surface movement.

2. Proposing a new method combining path planning and control concept to solve constrained trajectory planning based on the new model. A technical discussion is taken to compare with methods presented in the previous papers [16, 39, 40]. Convergence of the generalized trajectory is proved.
3. Three experiments are taken to prove the effectiveness of the proposed method for nonpoint obstacle avoidance in static and dynamic environments and cooperative dual-arm robot operation. The results show that most of the limitations of the original DMP presented by Pairet et al. [40] will be addressed.

6.2 Dynamic Movement Primitives and Constraints

This section contains two parts. The first one introduces the basic components for the DMP function proposed by Ijspeert et al. [6, 7], and the second part presents a general DMP model with classified constraints, which is the research foundation for the following section.

6.2.1 Dynamic Movement Primitives

The DMP model is first proposed in [7] as follows;

$$\begin{cases} \tau \dot{v} = \alpha_z \left(\beta_z (g - x) - v\right) + f(s) \\ \tau \dot{x} = v \end{cases}, \tag{6.1}$$

where $f(s) = \theta^T \Psi(s)$ is a linear combination of nonlinear radial basis functions $\varphi_i(s)$, and

$$\begin{aligned} theta &= [w_1, w_2, \ldots, w_n]^T, \\ \Psi(s) &= [\psi_1, \psi_2, \ldots, \psi_n]^T, \\ \psi_i &= \frac{\varphi_i(s) s}{\sum_{i=1}^{n} \varphi_i(s)}, \\ \varphi_i(s) &= \exp(-h_i (s - c_i)^2). \end{aligned} \tag{6.2}$$

where c_i and $h_i > 0$ are the centers and widths of $\varphi_i(s)$ respectively. The factors $\alpha_z, \beta_z >$ 0are coefficients of the linear part and has a unique attracting point at $x = g, v = 0$. $\tau > 0$ is a timing parameter adjusting speed before execution of movements. s is a phase variable to achieve the dependency of function $f(s)$ out of time.

Remark 6.1 The expression of the forcing function $f(s)$ is not unique. For example, Wang et al. [25] and Wu et al. [24] proposed the DMP plus method by adding a bias term to each kernel and used truncated kernels to achieve lower MSE in position deviation.

The dynamics of s is expressed by a canonical system

$$\tau\dot{s} = -\gamma s, \qquad \gamma > 0. \tag{6.3}$$

Equation (6.3) has an implicit relation with the time that we can modify the converging time by the factor γ. When $s \to 0$, as $t \to \infty$, the value of $f(s)$ trends to 0 and $[x, v]$ reaches to the stability point $[g, 0]$. The vector θ can be learned using supervised learning algorithms such as locally weighted regression (LWR) [17, 24, 25, 48–50]. The purpose of the calculating process is to minimize the error function:

$$\min \sum_{j=1}^{N} \left(f_j^{Tar}(s) - f_j(s)\right), \tag{6.4}$$

where $f_j(s)$ represents the item calculated by the jth trajectory in demonstration, and $f_j^{Tar}(s)$ is the target value of $f_j(s)$ as

$$f_j^{Tar}(s) = \tau\dot{v} - \alpha_z\left(\beta_z\left(g - x_j\right) - v_j\right). \tag{6.5}$$

6.2.2 *Generalization of Modified DMP and Constraints*

As mentioned in Sect. 6.1 can be improved for the situation with several constraints, we list some typical examples in Table 6.1. It is not hard to notice that both the basic expression of DMP and additional terms differ from each other with various purposes for manipulations. We express them with a general function:

$$\begin{cases} \dot{v} = f_2(g, x, v, \tau) + k(s, \tau) + \beta_g u(s, \tau) + \Delta u \\ \dot{x} = f_1(x, \tau, F) + g_1(x, \tau)v \\ \dot{s} = \Phi(s, \tau) \end{cases}, \tag{6.6}$$

where $\Phi(s, \tau)$ is the canonical system like (6.3) and $k(s, \tau)$ and $u(s, \tau)$ containing s. The difference of $k(s, \tau)$ and $u(s, \tau)$ is $k(s, \tau)$ represents a linear term like $k\,(g - x_0)\,s$ in [16] and $u(s, \tau)$ is the forcing function like $f(s)$ in [16, 20] etc. β_g is a constant factor and usually set as $\frac{1}{\tau}$. $f_2(g, x, v, \tau)$ is a general linear function of DMP like $\alpha_z\,(\beta_z(g - x) - v)$ in (6.1). Δu is an additional accelerating term caused by the constraints. All the DMP functions in Table 6.1 can obviously be abstracted into the formulation in (6.6). In the next section, a general method will be presented to deduce the exact expression of Δu.

Table 6.1 Typical improved DMP with constraints

Ref. No.	Constraints' expressions	Expressions of improved DMP
[16]	Obstacle constraint: $p(x, v) = \gamma R v \phi \exp(-\beta\phi)$, R is a rotation matrix with axis of $r = (o - x) \times v$ and o is the position of the obstacle, and $\phi = \cos^{-1}\left(\frac{(o-x)^T v}{\lvert o-x\rvert \cdot \lvert v\rvert}\right)$	$\tau\dot{v} = k(g - x) - d\dot{x} - k(g - x_0)s + kf(s)$ $+ p(x, v)$
[20]	Cooperative distance constraint: $c_i(x) = -\sum_{j\in N_i} \delta_{ij} \frac{x_i - x_j}{\lVert x_i - x_j\rVert} (\lVert x_i - x_j\rVert - d_{ij})$, d_{ij} is the desired distance between agent i and agent j;	$\tau\dot{v}_i = \alpha_z(\beta_z(g_i - x_i) - v_i) + f(s_i)$ $\tau\dot{x}_i = v_i + kc_i(x)$ $\tau\dot{s}_i = -\frac{\gamma_i s_i}{1 + \eta c_i^2(x)}$
[19]	Obstacle and internal force constraint: $C_{i,j} = cF_{i,j} l_{fi}$, $F_{i,j} = -$ $F_{j,i} = -k(d_d - d_a)$ represents the internal force between two agents, and d_d represents the desired distance between two effectors and d_a is the actual distance;	$\tau\dot{v}_i = k(d(g - x_i) - \dot{x}_i) + f(s) + c_2\dot{C}_{i,j}$ $\tau\dot{x}_i = v_i + C_{i,j}$ $C_{i,j} = cF_{i,j}(t)$
[17]	Curved surface constraint: Δa is added as the force coupling term related to the contact force error.	$\tau\dot{v} = k(d(g - x) - \dot{x}) + \alpha f(s, w) + \Delta a$ $\tau\dot{x} = v$

Following Table 6.1, we know that the constrained term usually has a referring point or trajectory like point o in [16], or x_j to x_i in [20], and position errors like $o - x$ or $\lVert x_i - x_j \rVert - d_{ij}$ are used to calculate the additional term. As trajectory points are x, we set the referring point of x as x^c, x_i^c and x_i are positions along the trajectories x^c and x. Additionally, we are inspired by the concept of "safety limits" in [50] and "safety margin" in [38] and propose h_i to describe the relations of x_i^c and x_i. All the possible relations can be shown as follows:

$$\underline{h}_i \le x_i^c - x_i \le \bar{h}_i \quad \text{or} \quad x_i^c - x_i \ge \bar{h}_i \quad \text{or/and} \quad x_i^c - x_i \le \underline{h}_i, \tag{6.7}$$

where $\bar{h}_i$ and $\underline{h}_i$ are upper and lower boundary functions, and $\bar{h}_i \ge h_i$. x_i^c and h_i (including $\bar{h}_i$ and $\underline{h}_i$) depend on operator's decisions or objective conditions learned on-time or by multi-disciplines in [39] and [40]. They also can be generated by local constraints and preferred routes. Different from learning methods of path planning, the proposed DMP model has two following advantages:

1. Ensuring the stability of trajectory. No matter the function in DMP except for the forcing term or the integrated BLFs method are proved to be stable and converged, but some neural network-based learning methods are not.

2. MPs can be integrated with intelligent learning methods and control theory to improve the learning efficiency and autonomy such as terms and can be manually set and learned by intelligent methods or calculated based on the analytic geometrical relations of interactive objects. After getting, the improved DMP model will be enacted to ensure convergence of the generalized trajectory satisfying inequalities in (6.7).

Furthermore, we divide (6.7) into two conditions, which both contain two cases.

Condition 6.1 Asymmetric constraints

- Case 1: $\underline{h}_i \leq x_i^c - x_i \leq \bar{h}_i$
- Case 2: $x_i^c - x_i \geq \bar{h}_i$ and/or $x_i^c - x_i \leq \underline{h}_i$

Condition 6.2 Symmetric constraints

- Case 1: $\left|x_i^c - x_i\right| \geq \bar{h}_i$
- Case 2: $\left|x_i^c - x_i\right| \leq \underline{h}_i$

Some assumptions are proposed as follows:

Assumption 6.1 The constrained trajectory is stable and satisfies if $s \to 0$, then

$$\dot{v}^c = f_2\left(g, x^c, v^c, \tau\right) \quad \text{and} \quad \left(x^c, v^c\right) \to (g, 0). \tag{6.8}$$

Assumption 6.2 The constrained trajectories and velocities are continuous and differentiable within every skill segmentation range.

Remark 6.2 Assumption 6.1 is the basis for trajectory convergence. According to [41], if $s \to 0$, the s-related term will decrease to 0 for normal DMP function. While for (6.6), $\dot{v} \to f_2(g, x, v, \tau) + \Delta u$ which means that the trajectory is influenced by error terms z_1, z_2. It is hoped that the additional constraints will not influence the trajectory converging to the target, thus the generated referring trajectory satisfying Assumption 6.1 ensures that $x \to x^c$ and $v^c, v \to 0$ at the destination.

Remark 6.3 Assumption 6.2 is proposed for calculation. Actually, a long-term demonstration will be segmented into several parts. The constraints may be presented discretely aside from the trajectory or even go across several segmented intervals. Two methods for generating continuous referring paths to connect the constraints are introduced in the next section, and Assumption 6.2 is the basis for the method.

6.3 Constrained Dynamic Movement Primitives

6.3.1 Constrained Referring Trajectory

As it is presented in Table 6.1, the referring variable x^c maybe a point, a surface, or a moving trajectory and its function is to guide x to satisfy conditions in (6.7), though some researchers proposed constrained dynamic movement primitives (CDMPs) that do not build a referring trajectory but use the transformed states instead of the original DMP states to maintain the joint trajectories within the safety limits [50].

Creating referring trajectory points x_i^c is the first step for the new DMP method. Here, we consider a case containing some local separate constraints and build a trajectory reserving the majority of the original path points and using the constraints to avoid obstacles or pass narrow avenues, then using the integral and continuous trajectory to guide robotic movements. Here, we provide two methods to achieve this purpose: interpolation and extended DMP method.

Set intervals divided by constraints and skill segmentations are $[x_0, x_1]$, $[x_2, x_3]$, ..., $\left[x_{i-1}, x_i\right]$. The interpolation method such as Lagrange interpolation polynomial provides some internal points to enrich the space between adjacent intervals, ensuring the continuity of positions and velocities. The extended The DMP method uses the generalized subskills to modify the interval goals, which is achieved in two steps.

1. Calculating original DMP within the interval $\left[x_{i-1}, x_i\right]$and the endpoint satisfying $g_i = x_i$;
2. Changing goal $g_i$to the start of new period x_{i+1}, then the current interval is extended from $\left[x_{i-1}, x_i\right]$to $\left[x_{i-1}, x_{i+1}\right]$.

The two methods are verified and compared in experiment 1. When the constraints come across each other, a common solution satisfying all the constraints will be selected. Moreover, we design the boundary functions according to manipulation requirements. Following convergence proof in Sect. 6.3.3, the trajectory will converge based on the condition of $z_1(s_i^{0+}) \in \left(\underline{h}_i, \bar{h}_i\right)$, where s_i^{0+}represents $s_i$in the ith interval at the start time $0+$. Then the junction point for the $(i-1)$thand i thintervals satisfies the condition of (6.7), such as for case 1, Condition 6.1, we have $\left(x_{i-1}^c - x_{i-1}\right)|_{s_{i-1}^{\infty}} = \left(x_i^c - x_i\right)|_{s_i^{0+}} \in \left(\underline{h}_{i-1}, \bar{h}_{i-1}\right)|_{s_{i-1}^{\infty}} \subseteq \left(\underline{h}_i, \bar{h}_i\right)|_{s_i^{0+}}$ to ensure continuity and avoid initial phase instability of path points.

6.3.2 Constrained Dynamic Movement Primitives

Barrier Lyapunov functions (BLF) are used for analyzing the stability of a closed-loop system, and they restrict full-state to the constraints [48, 51]. In this chapter, BLF are adopted to calculate the additional item of constrained DMP, enabling the trajectory calculated by constrained DMP to satisfy constraints in (6.7).

Define two new variables $z_1 = k_z\left(x_i^c - x_i\right), z_2 = d_z\left(\alpha_2 - v_i\right), k_z, d_z > 0$ are constant, and α_2 is a stabilizing function to be designed. Similar to asymmetric BLF candidate in [34], we taking case 1, Condition 6.1 as an example and build a Lyapunov function as

$$V^c = \frac{1}{2}\left(q(z_1)\log\frac{\bar{h}_i^2}{\bar{h}_i^2 - z_1^2} + (1 - q(z_1))\log\frac{\underline{h}_i^2}{\underline{h}_i^2 - z_1^2}\right) + \frac{1}{2}z_2^2, \tag{6.9}$$

where $q(\cdot) = 1$, if $\cdot > 0$ and $q(\cdot) = 0$, if $\cdot \le 0$. Then $\dot{V}^c$ is

$$\dot{V}^c = \left(\frac{q(z_1)}{\bar{h}_i^2 - z_1^2} + \frac{1 - q(z_1)}{\underline{h}_i^2 - z_1^2}\right) z_1 k_z\left(\dot{x}_i^c - \dot{x}_i\right) + z_2 d_z\left(\dot{\alpha}_2 - \dot{v}_i\right). \tag{6.10}$$

For realizing the stability Condition 6.1 of $\dot{V}^c \le 0$, the terms α_2, $u(s,\tau)$, and k_n are taken as

$$\begin{cases} \alpha_2 = \dfrac{d_z\left(\dot{x}_i^c - f_1(x,\tau,F)\right) + z_1}{d_z g_1(x,\tau)} \\ u(s,\tau) = \dfrac{\dot{\alpha}_2 - f_2(g,x,v,\tau) - k(s,\tau) - \Delta u + k_n z_2}{\beta_g} \\ k_n = \dfrac{k_2}{d_z} + \dfrac{k_z g_1(x,\tau)^2}{4d_z^2}\left(\dfrac{q(z_1)}{\bar{h}_i^2 - z_1^2} + \dfrac{1 - q(z_1)}{\underline{h}_i^2 - z_1^2}\right) \end{cases}, \tag{6.11}$$

where $k_2 > 0$ is a positive number and the proof is presented in Part C. Following (6.5), u^{Tar} without constraints is

$$u^{Tar} = \frac{\dot{v} - f_2(g,x,v,\tau) - k(s,\tau)}{\beta_g}. \tag{6.12}$$

It is desired that the additional constraints do not repeat the skill learning process (calculation of forcing functions), then the value error of $u(s,\tau) - u^{Tar}$ will decrease to zero, and Δu is

$$\Delta u = \dot{\alpha}_2 + k_n z_2 - \dot{v}. \tag{6.13}$$

Taking (6.13) into (6.6), we get the modified DMP satisfying constraints in case 1, Condition 6.1 in Sect. 6.2.2 as

$$\begin{cases} \dot{v} = f_2(g,x,v,\tau) + k(s,\tau) + \beta_g u(s,\tau) + \Delta u \\ \dot{x} = f_1(x,\tau,F) + g_1(x,\tau)v \\ \dot{s} = \Phi(s,\tau) \\ \Delta u = \dot{\alpha}_2 + k_n z_2 - \dot{v} \end{cases}. \tag{6.14}$$

Remark 6.4 For the regular DMP model like (6.1), the terms in (6.14) are instantiated as $f_1(x,\tau,F)=0$, $g_1(x,\tau)=\frac{1}{\tau}$, $f_2(g,x,v,\tau)=\frac{\alpha_z(\beta_z(g-x)-v)}{\tau}$, $k(s,\tau)=0$, $\beta_g=\frac{1}{\tau}$, then

$$\begin{cases} \tau\dot{v}=\alpha_z\left(\beta_z(g-x)-v\right)+f(s)+\tau\left(k_n z_2+\dfrac{\dot{z}_2}{d_z}\right) \\ \tau\dot{x}=v \\ \tau\dot{s}=-\gamma s \end{cases}. \tag{6.15}$$

Case 2: Similar to case 1, we build the BLFs candidate :

$$\begin{aligned} V^c=&\frac{1-q(z_1)}{2}\left((1-q(\underline{h}_i))\log\frac{z_1^2}{z_1^2-\underline{h}_i{}^2}+(1-q(\bar{h}_i))\log\frac{\bar{h}_i^2}{\bar{h}_i^2-z_1^2}\right) \\ &+\frac{q(z_1)}{2}\left(q(\underline{h}_i)\log\frac{\underline{h}_i{}^2}{\underline{h}_i{}^2-z_1^2}+q(\bar{h}_i)\log\frac{z_1^2}{z_1^2-\bar{h}_i^2}\right)+\frac{1}{2}z_2^2 \end{aligned}, \tag{6.16}$$

and the factors satisfying $\dot{V}^c\leq 0$ are calculated with different results in (6.11) except for the $u(s,\tau)$ as follows:

$$\begin{cases} \alpha_2=\dfrac{\dot{x}_i^c-f_1(x,\tau,F)}{g_1(x,\tau)}-\dfrac{K_2+K_1z_1^4}{d_zg_1(x,\tau)\left(K_2-K_1z_1^3\right)} \\ k_n=\dfrac{k_2}{d_z}+\dfrac{k_zg_1(x,\tau)^2}{4d_z^2}(K_1+K_2) \\ K_1=\dfrac{(1-q(z_1))\left(1-q(\bar{h}_i)\right)}{\bar{h}_i^2-z_1^2}+\dfrac{q(z_1)q(\underline{h}_i)}{\underline{h}_i{}^2-z_1^2} \\ K_2=\dfrac{\underline{h}_i{}^2(1-q(z_1))\left(1-q(\underline{h}_i)\right)}{z_1^2-\underline{h}_i{}^2}+\dfrac{\bar{h}_i^2q(z_1)q(\bar{h}_i)}{z_1^2-\bar{h}_i^2} \end{cases}. \tag{6.17}$$

Condition 6.3 Lyapunov functions are built with a different formation to (6.9), and V^c for cases 1 and 2 are defined as follows:

$$V^c=\frac{1}{2}\log\frac{{z_1}^2}{{z_1}^2-\bar{h}_i^2}+\frac{1}{2}z_2^2 \quad\text{and}\quad V^c=\frac{1}{2}\log\frac{\underline{h}_i{}^2}{\underline{h}_i{}^2-{z_1}^2}+\frac{1}{2}z_2^2. \tag{6.18}$$

Factors α_2, $u(s,\tau)$, and k_n enable the system stability condition $\dot{V}^c\leq 0$ to be satisfied are shown in Table 6.2.

Remark 6.5 Compared with (6.9), it is not hard to find that the expression of $u(s,\tau)$ does not change along with constraints, such that Δu is calculated with a fixed expression for any cases, though the other two factors α_2 and k_n will vary with different conditions.

Table 6.2 Symbol calculations for Condition 6.3

Symbol	$\lvert x_i^c - x_i\rvert \geq \bar{h}_i$	$\lvert x_i^c - x_i\rvert \leq \underline{h}_i$
α_2	$\frac{\dot{x}_i^c - f_1(x,\tau,F)}{g_1(x,\tau)} - \frac{1}{d_z g_1(x,\tau) z_1}$	$\frac{\dot{x}_i^c - f_1(x,\tau,F)}{g_1(x,\tau)} + \frac{z_1}{d_z g_1(x,\tau)}$
$u(s,\tau)$	$\frac{\dot{\alpha}_2 - f_2(g,x,v,\tau) - k(s,\tau) - \Delta u + k_n z_2}{\beta_g}$	
k_n	$\frac{k_2}{d_z} + \frac{k_z(\bar{h}_i g_1(x,\tau))^2}{4d_z^2(z_1^2 - \bar{h}_i^2)}$	$\frac{k_2}{d_z} + \frac{k_z(g_1(x,\tau))^2}{4d_z^2(\underline{h}_i^2 - z_1^2)}$

6.3.3 Convergence Proof

Some previous work integrated control methods with DMP and proved convergence of the generated DMP trajectory. In [41], Rai and Meier et al. built a Lyapunov stability criterion for DMP with a coupling term to prove the condition that if $s \to 0$ or $t \to \infty$, then $x \to g$, which means that the convergence of trajectory is changed by the additional term C^t. Similarly, for Δu in (6.14), we synthesize a Lyapunov function as follows:

$$V = V^1 + V^c = \frac{1}{2}(g-x)^T K(g-x) + \frac{1}{2}v^T v + V^c\left(x^c, x\right), \tag{6.19}$$

where $K = \alpha_z \beta_z > 0$, V^1 represents the two former terms of V. Taking case 1, Condition 6.1 as an example, to prove $\dot{V} < 0$, we first proves $\dot{V}^c \leq 0$. Taking (6.6) into (6.10), we have

$$\begin{aligned}
\dot{V}^c &= z_1 k_z \left(\frac{q(z_1)}{\bar{h}_i^2 - z_1^2} + \frac{1-q(z_1)}{\underline{h}_i{}^2 - z_1^2}\right)\left(\dot{x}_i^c - f_1(x,\tau,F) - g_1(x,\tau)\left(\alpha_2 - \frac{z_2}{d_z}\right)\right) \\
&\quad + d_z z_2\left(\dot{\alpha}_2 - f_2(g,x,v,\tau) - k(s,\tau) - \beta_g u(s,\tau) - \Delta u\right) \\
&= z_1 k_z \left(\frac{q(z_1)}{\bar{h}_i^2 - z_1^2} + \frac{1-q(z_1)}{\underline{h}_i{}^2 - z_1^2}\right)\left(\dot{x}_i^c - f_1(x,\tau,F) - g_1(x,\tau)\alpha_2 + \frac{z_1}{d_z}\right) \\
&\quad + d_z z_2 \left(\begin{array}{c} \dot{\alpha}_2 - f_2(g,x,v,\tau) - k(s,\tau) - \beta_g u(s,\tau) - \Delta u \\ + \left(\frac{k_2}{d_z} + \frac{k_z g_1(x,\tau)^2}{4d_z^2}\left(\frac{q(z_1)}{\bar{h}_i^2 - z_1^2} + \frac{1-q(z_1)}{\underline{h}_i{}^2 - z_1^2}\right)\right) z_2 \end{array}\right) \\
&\quad - \frac{k_z}{d_z}\left(\frac{q(z_1)}{\bar{h}_i^2 - z_1^2} + \frac{1-q(z_1)}{\underline{h}_i{}^2 - z_1^2}\right)\left(z_1 - \frac{g_1(x,\tau) z_2}{2}\right)^2 - z_2^T k_2 z_2
\end{aligned} \tag{6.20}$$

Using the symbols in (6.11), we have

$$\dot{V}^c = -\frac{k_z}{d_z}\left(\frac{q(z_1)}{\bar{h}_i^2 - z_1^2} + \frac{1-q(z_1)}{\underline{h}_i^2 - z_1^2}\right)\left(z_1 - \frac{g_1(x,\tau)z_2}{2}\right)^2 - z_2^T k_2 z_2. \quad (6.21)$$

Because $k_z, d_z, k_2 > 0$, $\bar{h}_i^2 - z_1^2 > 0$, and $\underline{h}_i^2 - z_1^2 > 0$, then if $z_1(s_i^{0+}) \in (\underline{h}_i, \bar{h}_i)$, where s_i^{0+} is the phase variable equals to the time $t = 0^+$ for the kth interval, then the inequality $\dot{V}^c \leq 0$ holds.

According to Remark 6.2, we know that when $s \to 0$, we have $\dot{V}^c(x^c, x) \leq 0$ and $z_1, z_2, \dot{z}_2 \to 0$, then (6.14) will degenerate to (6.22). Following [41], $V^1 < 0$ will be proved under the condition that if $s \to 0$, then $f(s) \to 0$. Then combining assumption 5.1, when $s \to 0$, we have $\tau\dot{v} = \alpha_z(\beta_z(g-x) - v)$. Setting $K = \alpha_z\beta_z$, we can get

$$\begin{aligned}\dot{V}^1 &= -\dot{x}^T K(g-x) + v^T\dot{v} \\ &= -\frac{v^T}{\tau}K(g-x) + \frac{\alpha_z}{\tau}v^T(\beta_z(g-x) - v). \\ &= -\frac{\alpha_z}{\tau}v^T v < 0\end{aligned} \quad (6.22)$$

It is easy to know that $\dot{V} \leq 0$ is achieved. The proof reveals that the additional term does not influence the system's stability.

6.4 Simulations and Experiments

In this section, we take two experiments with various constraints caused by human preferences, obstacles, and physics limitations to test the effectiveness of the propose methods. The first experiment is about static non-point obstacle avoid-ance to certify special zone avoidance and preferred route selection. The second experiment is about dual-arm robot cooperative operation with distance constraints of end effectors and joint links. The influence of measuring noise and errors is ignored compared with physical constraints. The first experiment uses a Omni joystick to record handling trajectories of the demonstrator and the latter one uses Kinect to track the skeleton data of human demonstration. We select DMP in (6.15) to generate trajectories with parameters $k_z = 150$, $d_z = 25$, $\alpha_z = 25$, $\beta_z = \frac{\alpha_z}{4}$, $\gamma = 0.7$, $k_2 = 10$ for all the cases, and other parameters are set differently for each experiment.

6.4.1 Experiment 1. Obstacle Avoidance

We first conduct a simulation in Experiment 1 with the setup shown in Fig. 6.1. The platform is built on a board with a 2D map which is a square of 20×16 cm

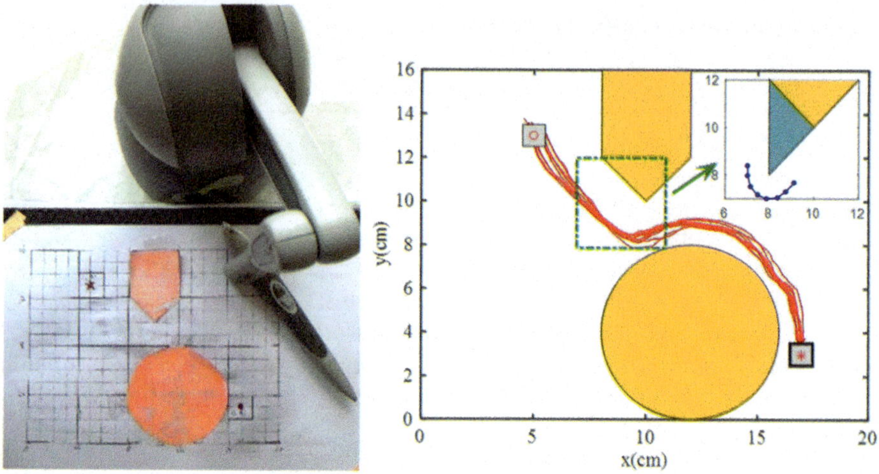

(a) (b)

Fig. 6.1 Experimental setups and conditions for static obstacle avoidance **a** experiment equipment and **b** simulation environment equivalent to the physical: orange blocks are original obstacles and the triangular blue block is an additional obstacle and the blue arc is the prospected trajectory with a safe distance to the new obstacle to protect its corner. Human-demonstrated motions are colored in red and have no conflict with the original obstacles

and the start point is located at [3, 16] and endpoint at [5, 14]. The values are recorded and presented in the virtual environment of MATLAB in Fig. 6.1b. Here we set predesigned preferred trajectory points $x_i^c \in \mathbb{R}^{2\times 1}$ (blue line with dots in the zoomed subfigure) to avoid confliction with the additional block (blue triangle area). Furthermore, we set a "safety margin" $h_i = 0.5$ to limit the trajectory ranges that satisfy:

$$\left\|x_i^c - x_i\right\|_2 \leq h_i, \quad x_i \in \mathbb{R}^{2\times 1}. \tag{6.23}$$

Figure 6.2 shows the generated constrained trajectories and simulation results. Using the BP-AR-HMM method mentioned in [27], demonstrations are segmented and generalized into two sub-skills (red and blue lines in (Fig. 6.2a), each of which is reshaped with 100 points. Due to the new obstacle, the robot following the path will encounter a collision. Furthermore, the constraints only affect the second sub-skill points, judging by the coordinate ranges in the X-axis. Therefore, we redivide the ranges of the second sub-skill into three regions (two black and a red line) and generate the whole-range potential trajectory points by the two methods developed in Sect. 6.3.1.

Figure 6.2b shows a referring path with a constrained path point generated by cubic spline interpolation that reserves part of the results of the original DMP. It connects original path point sets and constrained paths with optimized interpolation points. The trajectories generated by the extended DMP model, shown in (Fig. 6.2c), come across the original obstacles and are not suitable for the task in this chapter.

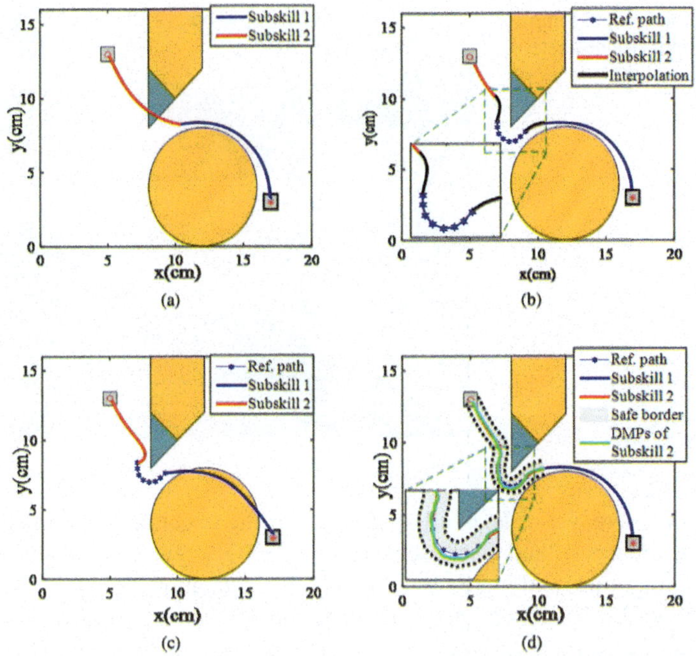

Fig. 6.2 Experimental results of static obstacle avoidance **a** Trajectories of the original DMP model, containing two segmentations: blue line for the first and red line for the second; **b** Constrained path generated by the interpolation method: the thin blue line with dot is the constrained path, the blue line is the trajectory of subskill 1, and the red line is the reserved original subskill 2, and the black line is the interpolation trajectory; **c** Constrained path generated by extended DMP; the thin blue line with dot is the constrained path, and the blue and red lines are the extended DMP trajectories to connect the constrained path; **d** Final generated trajectories based on the constrained route in subfigure **b**, compared with (b), the additional area and lines are: the green line is the generated 2nd DMP subskill and the gray area is the safe margin to protect collision with obstacles

The main reason is environmental factors (shape of obstacles and conflict, etc.) are not considered in the original DMP function. Thus, we select Fig. 6.2b as the referring path in this chapter. But, it does not mean that which of the two methods is better but depends on operational demands for different tasks. Usually, for a short-distance point-to-point connection, the interpolation method is easier to use, and it will not change the previously planned path, while the DMP-based method changes the distinction of the trajectory segment, which can be generalized in the space and time scales. The obstacles and other limitations should be considered during the design process for both cases.

Finally, Fig. 6.2d shows the final path generated by the modified DMP. Under the conditions in (6.23), it researches to the largest distance to the constrained path, but still keeps within a certain distance to x_2^c and converges to the constrained path quickly at the end of arch points.

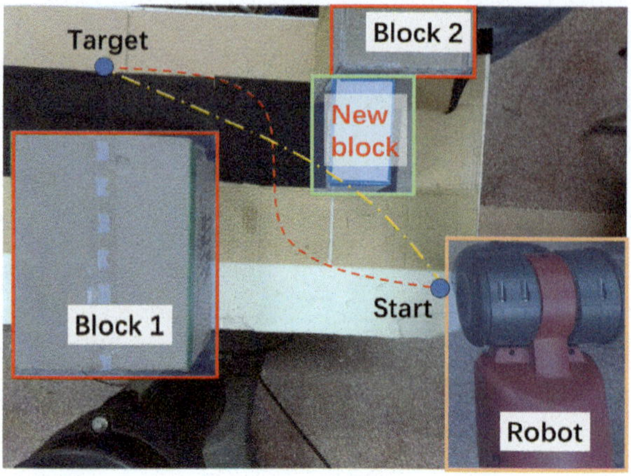

Fig. 6.3 Experimental setup for single robot arm obstacle avoidance

Remark 6.6 Here, we make a further analysis of the selection of parameters. Actually, the parameters α_z and β_z are selected as $\beta_z = \frac{\alpha_z}{4}$ to guarantee that the globally stable system in (6.1) and (6.15) are critically damped, allowing x to monotonically converge to g [25]. Factors k_z and d_z relate to the position and velocity errors, and a similar rate to x and v is selected as $k_z = \frac{d_z^2}{4}$. Here, we select an approximate solution as $k_z = 150$ and $d_z = 25$. The simulation shows that k_2 realizes system stability and influences the dynamic performance of the generated trajectory. A larger k_2 will achieve a larger tracking error of $x - x_i^c$ for this experiment, but it is not a conclusive result for all cases.

Then we conduct an experiment using the Baxter robot for obstacle avoid-ance based on the simulation results with the following setup in Fig. 6.3. There are two obstacles along the way from the start to the target. We first make a demonstration first as the trajectory in the same way as in Fig. 6.1 and the generalized is shown as the yellow dot-dash line in Fig. 6.3. Then we add a new obstacle along the generalized pathway, therefore a new trajectory needs to be generated to avoid the newly added block as the red dashed line shown in Fig. 6.3.

Figures 6.4 and 6.5 present two comparative experiments for obstacle avoidance using the original DMP and the improved DMP. It is evident that the generalized trajectory cannot traverse the newly added obstacle due to the limitations of the DMP, which leads to uncontrollable internal trajectory points between the start and end points. In contrast, the improved method takes into account the constraints that could affect the internal trajectory points and effectively handles unseen scenarios.

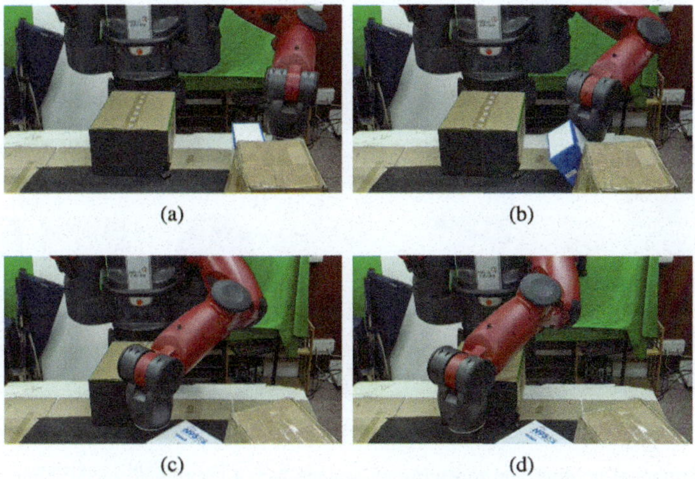

Fig. 6.4 Obstacle avoidance experiment using the original DMP

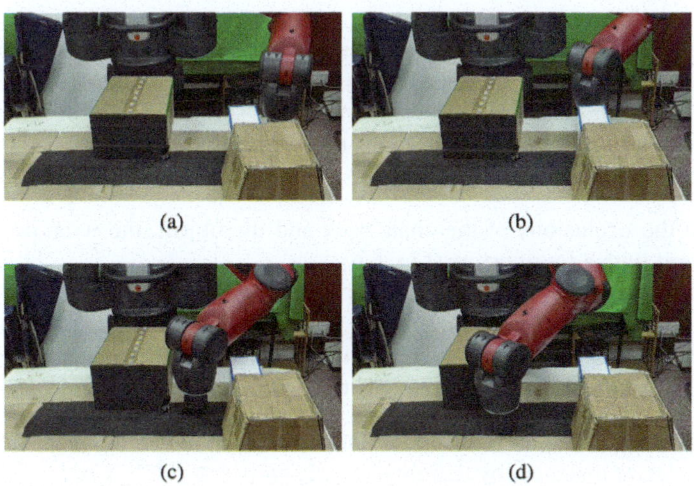

Fig. 6.5 Obstacle avoidance experiment using the improved DMP

6.4.2 Experiment 2. Bimanual Cooperative Manipulation

DMP are widely used for multi-joint humanoid skill learning such as object-lifting and cutting food [52–56]. Most of the papers only learned and generalized joint information [16, 17, 19, 24–27, 33, 37, 52], while for cooperative bimanual manipulation with a strict distance limitation of robot end effectors, e.g., holding a box or grasping a bar, only training data for joint information is not enough. Thus, cooperative DMP [43–45] usually studied the trajectories of robot ends in the Cartesian coordinate. In

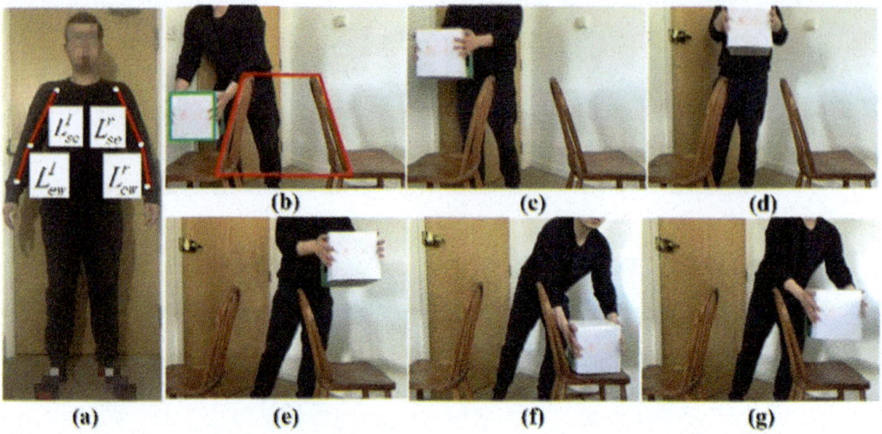

Fig. 6.6 Bimanual holding and replacing an opaque box **a** coordinate of the human body and related links length and **b–f** moving steps of the object

this experiment, we combine these two cases and consider the constraints both in the joint distance and relative distance of robot ends. First, we build a data set about the positions of the shoulders, elbows, and wrists (for simplification, we use the wrists data instead of hands) of demonstrations measured by Kinect. There are two major problems with the measured data:

1. Due to the occlusions of the chair back and the object, the wrist positions are measured with a lot of errors (Fig. 6.6).
2. Compared with the wrists and the elbows, the data of the shoulders are more stable and credible. As is shown in Fig. 6.7, we calculate the internal distances of the wrists, elbows, and shoulders to middle values of ten times demonstrations. It can be seen that the data width of the wrists is about twice of the shoulders.

Because the cooperative manipulation of robot ends endures the constraint of constant relative distance, by defining x^w_j, x^e_j, and $x^s_j \in \mathbb{R}^{3\times 1}$, $j = l, r$ as positions of the wrists, elbows, and shoulders, we proposed the following constraints for robot end effectors:

$$wrist: \begin{cases} \left\| x^{dw}_j - x^w_j \right\|_2 \le h^w_j \\ x^{dw}_j = x^w_j + d^w_j \end{cases}, j = l, r, \tag{6.24}$$

where $d^w_j \in \mathbb{R}^{3\times 1}$ represents the desired relative distance, and $\left| d^w_j \right|$ is a constant, and h^w_j is an acceptable distance calculation error caused by object deformation or measuring errors. Define L^j_{se}, L^j_{ew} as the vectors between the shoulders and the elbows, and the elbows and the wrists (Fig. 6.6a), the constraints of other linked joints are:

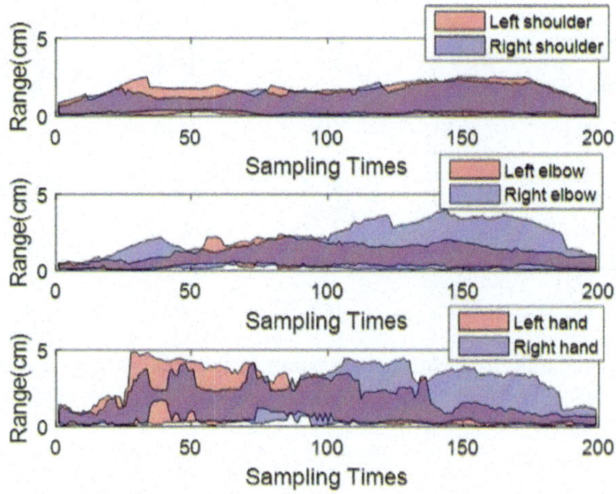

Fig. 6.7 Spreading distances of the wrist, elbow, and shoulder. In each subfigure, the red area represents the distance range of the left hand, and the blue area represents the results of the right hand, and the purple area represents the intersection area

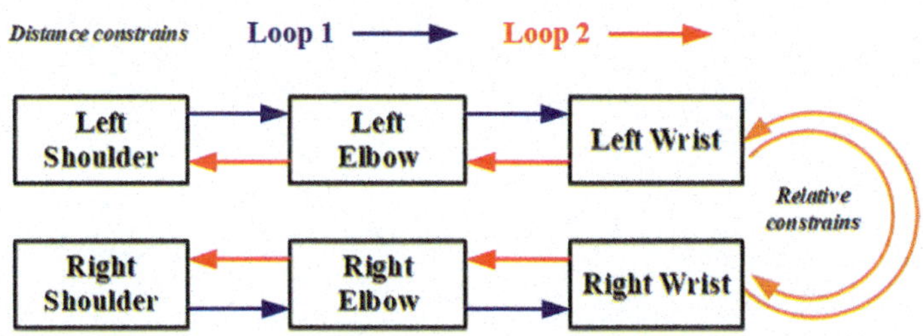

Fig. 6.8 Calculation diagram for cooperative DMP that contain two loops of distances constraints and one relative constraint

$$shoulder : \begin{cases} \|x_j^{ds} - x_j^s\|_2 \leq h_j^s \\ x_j^{ds} = L_{se}^j + x_j^e \end{cases}, elbow : \begin{cases} \|x_j^{de} - x_j^e\|_2 \leq h_j^e \\ x_j^{de} = L_{ew}^j + x_j^w \end{cases}, j = l, r. \tag{6.25}$$

Since the mentioned relative distance constraints (6.24) and distance constraints (6.25) are contradicted-every joint is affected by relative constraints, but the position error rates of the end effectors are higher than the other joints, we propose the calculation diagram in Fig. 6.8.

The distance constraints are utilized twice: in loop 1, we use the shoulder and elbow data to correct the wrist data with errors. Then, the relative positions of the wrists are revised again by (6.24). Finally, using inverse calculation (from endpoint to the former two joints, loop 2), we amend the shoulder and elbow data under the distance constraints (6.25).

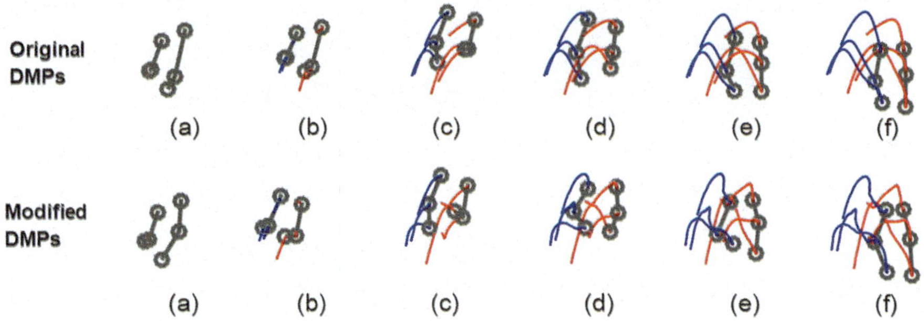

Fig. 6.9 Positions and trajectories of skeleton points in cooperative DMP corresponding with the operation process (see **a**–**f**). The blue lines are generated skills of the left shoulder, elbow, and wrist. The red lines are the results of the right armlabel

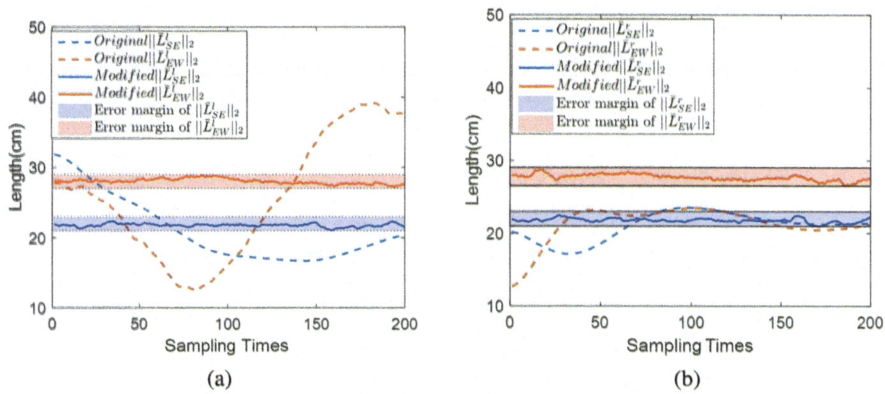

Fig. 6.10 Distance between adjacent joints (shoulder and elbow, elbow and wrist) **a** results of the left arm and **b** Results of the right arm. The solid lines of the two figures are the results of improved DMP, and the dashed lines are the results of original DMP with the same dataset, and the colored areas are designed error margins of the generated link lengths. The generated distance within the area means the constrained condition Eq. (6.21) is satisfied

The joint distance in Fig. 6.6a is measured with $\|L_{se}\|_2 = 22\,\text{cm},$, $\|L_{ew}\|_2 = 28\,\text{cm}$, and $\left\|d_j^w\right\|_2 = 33\,\text{cm}$ is the length of the box, and $h_j^s = h_j^e = h_j^w = 1\,\text{cm},\ j = l, r$ are acceptable trajectory tracking errors and object deformation degree. The simulation results are presented in Figs. 6.9, 6.10 and 6.11.

Figure 6.9 shows the trajectories generated by constrained DMP are different from original DMP, and data analysis is presented in Figs.6.10 and 6.11. Figure 6.10a shows the distance between adjacent joints of the left arm. It can be seen that the real joint distance of $\left\|\bar{L}_{se}^l\right\|_2$ and $\left\|\bar{L}_{ew}^l\right\|_2$ calculated by the original DMP vary within ranges of [12, 52] and [25, 39], which are influenced seriously by the measuring noises and errors and can't meet the actual situation. So as the results of $\left\|\bar{L}_{se}^r\right\|_2$ and $\left\|\bar{L}_{ew}^r\right\|_2$ of the right arm in Fig. 6.10b. The trajectories of the modified DMP keep a

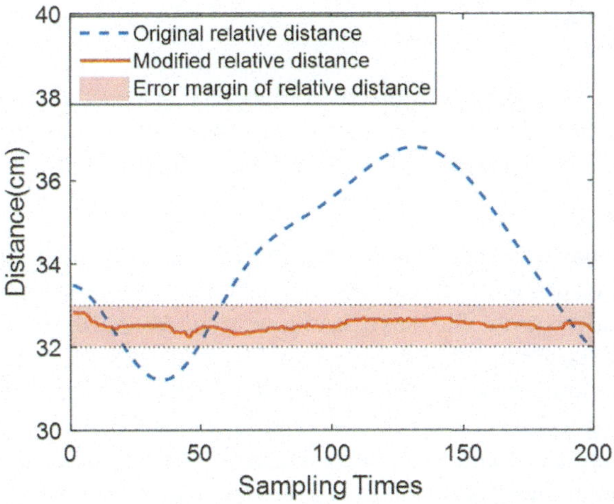

Fig. 6.11 Relative distance of the left hand and right hand in the generated DMP path (red line is the results of modified DMP, and blue line is the results of original DMP). The red area is the error margin of the relative distance introduced in (6.24)

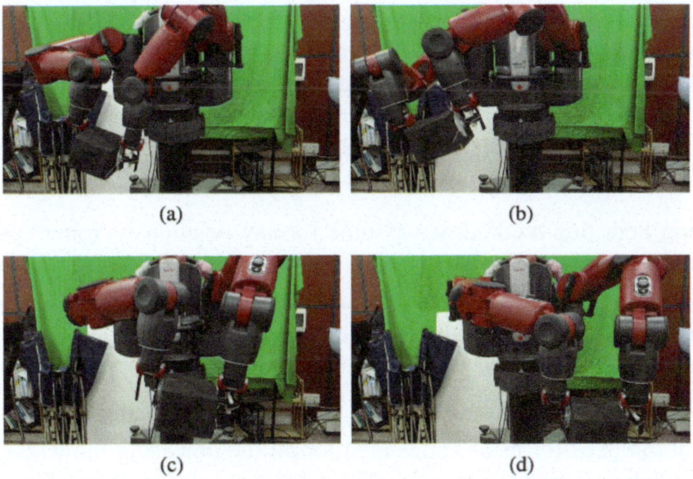

Fig. 6.12 Experiment for dual-arm manipulation using the improved DMP

stable distance as predesigned in (6.25), which presents that the influence of noise is reduced. Figure 6.11 shows the relative distance of the wrist (hands). The modified DMP framework decreases the varying range from [34, 57] to [24, 25], ensuring bimanual manipulation stability (Fig. 6.12).

6.5 Discussion

In this section, we will make a discussion about if the method can address most of the problems stated by Pairet et al. in [40]. Here, we classify the problems into the following three categories and most of them are proved through experiments in Sect. 6.4.

1. **Dead-zone and preferred route**: In [40], the authors stated that for the point obstacle avoidance method like [16], the analytical term $\dot{\theta} = \gamma\theta exp(-\beta\theta)$ (β is a constant and θ is a variable) has a dead zone around 0, and the system will become less reactive as the heading towards the obstacle narrows, thus compromising the method's reliability. But for the proposed method, we can select multi-piece trajectories or make a tiny modification of the planned path around the dead zone as new references, and the preferred route also can be designed flexibly following the operator's preference.
2. **Moving non-point obstacles**: Some papers used the DMP model for the case with moving obstacles, such as Hoffmann et al. [16]. studied the robustness of dynamic systems against perturbations for obstacle avoidance and extended it to cases with multiple obstacles and moving obstacles. Park et al. [39] designed a special dynamic potential function for the moving obstacles. But, these methods don't suit non-point obstacle cases. As some existed research addressed the object detection and real-time trajectory navigation based on autoregressive model [57], Kalman filter or tracking method [58] to solve the problem for non-demonstration cases, we can select a suitable method as prepositive treatment of our proposed DMP framework, but it is not the main work in this chapter.

By using detection method, we assume for any position x_i, robot can perceive edges and forecast dynamic positions of the obstacle within the next few steps as $X^o = \left[x_k^o, x_{k+1}^o, \ldots, x_{k+N}^o\right]^T$ and x_k^o is a data set filled with detected obstacle positions. Then path point x_k^c and constraints are generated by setting a 'safety margin' [38] toward the obstacle. The advantages of combining the path prospective method and the proposed DMP model are:

1. Without the need for prior knowledge for all the objects in the whole map and predesigned trajectories e.g., [39]. The robot can select a planned path to avoid conflicts and return to the original DMP route once facing obstacles.
2. Useful for multi-segment convex obstacle avoidance. The referring path connects several local constraints to form a continuous long trajectory, and most of the constraints can be designed with different conditions change in dynamic environment timely and flexibly.
3. Time-consuming and noise: Actually, the calculation in (6.14) is more complex than the methods in [39] and [16]. However, the proposed method just modifies trajectory points along part of the previous path, which can avoid extra calculation for the whole trajectory. For the influence of measuring noise, it is hard to diminish but we can estimate the boundaries of noise varying range and add them for

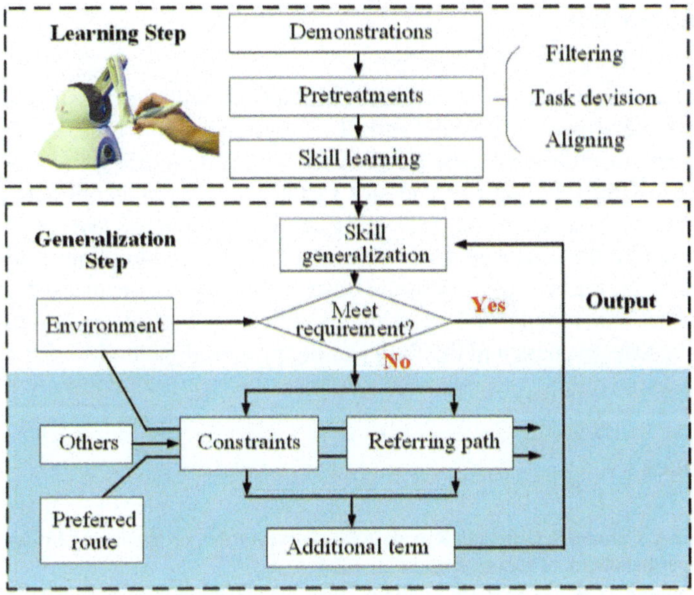

Fig. 6.13 Calculation procedures of the constrained DMP framework

the design of h_i. For example, in experiment 3, using physical constraints (e.g., distance of joint links and end effectors), the measuring noises and errors will be filtered and amended. Though the proposed method solves most problems with universal equations for various cases, similar to most of the path replanning methods, it depends on prediction accuracy of the environmental parameters and referring trajectories that can be calculated timely.

Another issue for discussion is the calculation procedures (Fig. 6.13) of the improved DMP method in the calculation.

Similar to the normal skill learning process based on DMP, there are two steps from demonstrations to applications in new cases. After pretreatments such as data filtering, task division and data aligning etc., the forcing function will be learned and then the skill (trajectory) will be generalized based on the given new start and end point and timing factor. But, due to the new cases occurring in the environment, the primitive generated path may not satisfy the operational demands. Following the changes in the environment, the preferred routes of the operators and some physical limitations will be used to generate the referring path (Sect. 6.3.1) with classified constraints (Sect. 6.2.2). Using (6.11) and (6.13), the additional term Δu will be calculated and added to DMP functions and the new trajectory will be generalized. The term Δu does not influence the skill learning results and is easy to be embedded in the original DMP function. The chosen parameters such as τ, α_z and β_z are the same as in (6.4) and (6.5). Next, we will follow the diagram to explore the potential applications of the method for different tasks with various conditions.

6.6 Conclusion

In this chapter, we propose a general DMP framework with the classified constraints based on a generalized DMP model and BLFs. Compared with other improved DMP models, a general additional term is calculated with certain expressions to encounter the changes of the environment and suit various tasks such as obstacle avoidance and cooperative manipulation by selecting suitable parameters. We make a further discussion about the advantages of this method such as addressing dead-zone and human online preferred route, non-point obstacle, and part of the measuring noise, as well as limitations. Finally, two experiments with some typical constraints are taken to verify the effectiveness and flexibility of the proposed method.

References

1. Billard A, Calinon S, Dillmann R et al (2008) Survey: robot programming by demonstration. Springer handbook of robotics, pp 1371–1394
2. Tanwani AK, Calinon S (2016) Learning robot manipulation tasks with task-parameterized semitied hidden semi-Markov model. IEEE Rob Autom Lett 1(1):235–242
3. Khansari-Zadeh SM, Billard A (2011) Learning stable nonlinear dynamical systems with Gaussian mixture models. IEEE Trans Rob 27(5):943–957
4. Calinon S, Sauser E, Caldwell ED et al (2009) A probabilistic approach based on dynamical systems to learn and reproduce gestures by imitation. IEEE Rob Autom Mag 17(2):44–54
5. Ng AY, Russell SJ (2000) Algorithms for inverse reinforcement learning. Paper presented at the 2000 Proceedings of the seventeenth international conference on machine learning, Stanford University, Standord, CA, USA, June 29–July 2, 2000
6. Ijspeert AJ, Nakanishi J, Hoffmann H, Pastor P, Schaal S (2013) Dynamical movement primitives: learning attractor models for motor behaviors. Neural Comput 25(2):328–373
7. Ijspeert AJ, Nakanishi J, Schaal S (2001) Trajectory formation for imitation with nonlinear dynamical systems. Paper presented at the proceedings 2001 IEEE/RSJ international conference on intelligent robots and systems. Expanding the societal role of robotics in the the next millennium (Cat. No.01CH37180), Maui, HI, USA, 29 October 2001–03 November 2001
8. Ude A, Gams A, Asfour T et al (2010) Task-specific generalization of discrete and periodic dynamic movement primitives. IEEE Trans Rob 26(5):800–815
9. Li Z, Zhao T, Chen F, Hu Y, Su CY, Fukuda T (2017) Reinforcement learning of manipulation and grasping using dynamical movement primitives for a humanoid like mobile manipulator. IEEE/ASME Trans Mechatron 23(1):121–131
10. Colomé A, Torras C (2018) Dimensionality reduction for dynamic movement primitives and application to bimanual manipulation of clothes. IEEE Trans Rob 34(3):602–615
11. Calinon S, Billard A (2009) Statistical learning by imitation of competing constraints in joint space and task space. Adv Rob 23(15):2059–2076
12. Rozo L, Calinon S, Caldwell DG (2014) Learning force and position constraints in human-robot cooperative transportation. Paper presented at the The 23rd IEEE international symposium on robot and human interactive communication, Edinburgh, UK, 25–29 August 2014
13. Yang C, Chen C, He W, Cui R, Li Z (2018) Robot learning system based on adaptive neural control and dynamic movement primitives. IEEE Trans Neural Netw Learn Syst 30(3):777–787
14. Yang C, Chen C, Wang N, Ju Z, Fu J, Wang M (2018) Biologically inspired motion modeling and neural control for robot learning from demonstrations. IEEE Trans Cogn Dev Syst 11(2):281–291

15. Abu-Dakka FJ, Kyrki V (2020) Geometry-aware dynamic movement primitives. Paper presented at the 2020 IEEE international conference on robotics and automation (ICRA), Paris, France, 31 May 2020–31 August 2020
16. Hoffmann H, Pastor P, Park DH, Schaal S (2009) Biologically-inspired dynamical systems for movement generation: automatic real-time goal adaptation and obstacle avoidance. Paper presented at the 2009 IEEE international conference on robotics and automation, Kobe, Japan, 12–17 May 2009
17. Han L, Kang P, Chen Y, Xu W, Li B (2019) Trajectory optimization and force control with modified dynamic movement primitives under curved surface constraints. Paper presented at the 2019 IEEE international conference on robotics and biomimetics (ROBIO), Dali, China, 06–08 December 2019
18. Lu Z, Wang N, Yang C (2022) A novel iterative identification based on the optimised topology for common state monitoring in wireless sensor networks. Int J Syst Sci 53(1):25–39
19. Gams A, Nemec B, Ijspeert AJ, Ude A (2014) Coupling movement primitives: Interaction with the environment and bimanual tasks. IEEE Trans Rob 30(4):816–830
20. Umlauft J, Sieber D, Hirche S (2014) Dynamic movement primitives for cooperative manipulation and synchronized motions. Paper presented at the 2014 IEEE international conference on robotics and automation (ICRA), Hong Kong, China, 31 May 2014–07 June 2014
21. Huang R, Cheng H, Qiu J, Zhang J (2019) Learning physical human-robot interaction with coupled cooperative primitives for a lower exoskeleton. IEEE Trans Autom Sci Eng 16(4):1566–1574
22. Saveriano M, Lee D (2018) Incremental skill learning of stable dynamical systems. Paper presented at the 2018 IEEE/RSJ international conference on intelligent robots and systems (IROS), Madrid, Spain, 01–05 October 2018
23. Lemme A, Reinhart RF, Steil JJ (2014) Self-supervised bootstrapping of a movement primitive library from complex trajectories. Paper presented at the 2014 IEEE-RAS international conference on humanoid robots, Madrid, Spain, 18–20 November 2014
24. Wu Y, Wang R, D'Haro LF, Banchs RE, Tee KP (2018) Multi-modal robot apprenticeship: imitation learning using linearly decayed DMP+ in a human-robot dialogue system. Paper presented at the 2018 IEEE/RSJ international conference on intelligent robots and systems (IROS), Madrid, Spain, 01–05 October 2018
25. Wang R, Wu Y, Chan WL, Tee KP (2016) Dynamic movement primitives plus: for enhanced reproduction quality and efficient trajectory modification using truncated kernels and local biases. Paper presented at the 2016 IEEE/RSJ international conference on intelligent robots and systems (IROS), Daejeon, 09–14 October 2016
26. Schaal S (2006) Dynamic movement primitives—a framework for motor control in humans and humanoid robotics. In: Adaptive motion of animals and machines. Springer, Tokyo
27. Yang C, Zeng C, Cong Y, Wang N, Wang M (2018) A learning framework of adaptive manipulative skills from human to robot. IEEE Trans Indus Inf 15(2):1153–1161
28. Wang N, Chen C, Yang C (2020) A robot learning framework based on adaptive admittance control and generalizable motion modeling with neural network controller. Neurocomputing 390:260–267
29. Liu Z, Lin W, Yu X, Rodriguez-Andina JJ, Gao H (2021) Approximation-free robust synchronization control for dual linear motors driven systems with uncertainties and disturbances. IEEE Trans Indus Electron 69(10):10500–10509
30. Shi P, Sun W, Yang X, Rudas I, Gao H (2022) Master-slave synchronous control of dual drive gantry stage with cogging force compensation. IEEE Trans Syst Man Cybern Syst 53(1):216–225
31. Zhang Y, Li M, Yang C (2021) Robot learning system based on dynamic movement primitives and neural network. Neurocomputing 451:205–214
32. Si W, Wang N, Yang C (2023) Composite dynamic movement primitives based on neural networks for human-robot skill transfer. Neural Comput Appl 35(32):23283–23293
33. Calinon S (2020) Gaussians on Riemannian manifolds: applications for robot learning and adaptive control. IEEE Rob Autom Mag 27(2):33–45

34. Huang H, Zhang T, Yang C, Chen CP (2019) Motor learning and generalization using broad learning adaptive neural control. IEEE Trans Indus Electron 67(10):8608–8617
35. Krug R, Dimitrov D (2015) Model predictive motion control based on generalized dynamical movement primitives. J Intell Rob Syst 77(1):17–35
36. Matsubara T, Hyon SH, and Morimoto J (2010) Learning stylistic dynamic movement primitives from multiple demonstrations. Paper presented at the 2010 IEEE/RSJ international conference on intelligent robots and systems, Taipei, Taiwan, 18–22 October 2010
37. Akbulut M, Oztop E, Seker MY, Hh X, Tekden A and Ugur E (2021) Acnmp: skill transfer and task extrapolation through learning from demonstration and reinforcement learning via representation sharing. Paper presented at the proceedings of the 2020 conference on robot learning, PMLR, virtual, 16–18 November 2020
38. Khansari-Zadeh SM, Billard A (2012) A dynamical system approach to realtime obstacle avoidance. Autonom Rob 32:433–454
39. Park DH, Hoffmann H, Pastor P et al (2008) Movement reproduction and obstacle avoidance with dynamic movement primitives and potential fields. Paper presented at the humanoids 2008—8th IEEE-RAS international conference on humanoid robots, Daejeon, Korea (South), 01–03 December 2008
40. Pairet È, Ardón P, Mistry M et al (2019) Learning generalizable coupling terms for obstacle avoidance via low-dimensional geometric descriptors. IEEE Rob Autom Lett 4(4):3979–3986
41. Rai A, Meier F, Ijspeert A et al (2014) Learning coupling terms for obstacle avoidance. Paper presented at the 2014 IEEE-RAS international conference on humanoid robots, Madrid, Spain, 18–20 November 2014
42. Umlauft J, Sieber D, Hirche S (2014) Dynamic movement primitives for cooperative manipulation and synchronized motions. Paper presented at the 2014 IEEE international conference on robotics and automation (ICRA), Hong Kong, China, 31 May 2014–07 June 2014
43. Rozo L, Calinon S, Caldwell DG et al (2016) Learning physical collaborative robot behaviors from human demonstrations. IEEE Trans Rob 32(3):513–527
44. Kulvicius T, Biehl M, Aein MJ et al (2013) Interaction learning for dynamic movement primitives used in cooperative robotic tasks. Rob Autonom Syst 61(12):1450–1459
45. Solak G, Jamone L (2019) Learning by demonstration and robust control of dexterous in-hand robotic manipulation skills. Paper presented at the 2019 IEEE/RSJ international conference on intelligent robots and systems (IROS), Macau, China, 03–08 November 2019
46. Lee H, Kim H, Kim HJ (2018) Planning and control for collision-free cooperative aerial transportation. IEEE Trans Autom Sci Eng 15(1):189–201
47. Qiao H, Wang M, Su JH et al (2014) The concept of 'attractive region in environment' and its application in high-precision tasks with low-precision systems. IEEE/ASME Trans Mechatron 20(5):2311–2327
48. Yang C, Wang X, Cheng L et al (2016) Neural-learning based telerobot control with guaranteed performance. IEEE Trans Cybern 47(10):3148–3159
49. Huang H, Yang C, Chen CLP (2020) Optimal robot-environment interaction under broad fuzzy neural adaptive control. IEEE Trans Cybern 51(7):3824–3835
50. Duan A, Camoriano R, Ferigo D et al (2018) Constrained DMPs for feasible skill learning on humanoid robots. Paper presented at the 2018 IEEE-RAS 18th international conference on humanoid robots (humanoids), Beijing, China, 06–09 November 2018
51. He W, Chen Y, Yin Z (2016) Adaptive neural network control of an uncertain robot with full-state constraints. IEEE Trans Cybern 46(3):620–629
52. Ogrinc M, Gams A, Petrič T et al (2013) Motion capture and reinforcement learning of dynamically stable humanoid movement primitives. Paper presented at the 2013 IEEE international conference on robotics and automation, Karlsruhe, Germany, 06–10 May 2013
53. Elbasiony R, Walid G (2018) Humanoids skill learning based on real-time human motion imitation using kinect. Intell Serv Rob 11:149–169
54. Yang C, Zeng C, Fang C et al (2018) A DMPs-based framework for robot learning and generalization of humanlike variable impedance skills. IEEE/ASME Trans Mechatron 23(3):1193–1203

55. Yang C, Zeng C, Cong Y et al (2018) A learning framework of adaptive manipulative skills from human to robot. IEEE Trans Indus Inf 15(2):1153–1161
56. Zeng C, Yang C, Zhong J et al (2019) Encoding multiple sensor data for robotic learning skills from multimodal demonstration. IEEE Access 7:145604–145613
57. Elnagar A, Gupta K (1998) Motion prediction of moving objects based on autoregressive model. IEEE Trans Syst Man Cybern Part A Syst Humans 28(6):803–810
58. Fu J, Chai T, Su CH et al (2013) Motion/force tracking control of nonholonomic mechanical systems via combining cascaded design and backstepping. Automatica 49(12):3682–3686

Chapter 7
Incremental Motor Skill Learning and Generalization from Human Dynamic Reactions Based on Dynamic Movement Primitive and Fuzzy Logic System

7.1 Introduction

Learning from demonstrations (LfD) provides a low-cost and effective way for robot programming and has drawn a lot of research attention in recent decades [1]. Dynamic movement primitives (DMP), proposed by Schaal [2], is a practical branch of LfD and first enables an artificial agent to act a complex human-like action in a versatile and creative manner [3]. As is mentioned in [3], DMP is full of advantages, such as simple and elegant formulation, convergence to a given target, flexibility for complex behaviours, and the capability of reacting to some external perturbations. DMP is then developed into many algorithms, such as orientation DMP, and rotation matrix DMP, and is widely used for cooperative manipulation [4, 5], obstacle avoidance [6, 7, 11–16], and variable impedance control [17, 18].

As seen from previous research, most published DMP-related methods extracted skills from single or multiple demonstrations performed in the same task, which are named ordinary skills in this article. In the case exhibited in Fig. 7.1, a person picks up a cup from the start position and takes it to the destination following the red path. When a human faces a sudden obstacle or danger, his or her muscles will make a nervous reaction to avoid conflict. This muscle shrink reaction, defined as the reactive skill in this article, has seldom been discussed in previous robot skill learning research. This reactive skill has two characteristics: First, it is an additional skill, in that if there is no accident, it will not be activated, and the human will move as originally planned. If an accident occurs, the reactive skill will act to change the ordinary path. Second, different people react differently. Humans will follow their accustomed way of doing business. Therefore, this article aims to propose a method for realizing both ordinary and characteristic reactive skills learning and generalization.

First, we will compare this article issue with related research topics on DMP-based obstacle avoidance and incremental skill learning. Obstacle avoidance is a typical application of DMP demonstrated by adding a potential field term to modify

C. Yang et al., *Advanced Teleoperation and Robot Learning for Dexterous Manipulation*, Springer Tracts in Advanced Robotics 160,
https://doi.org/10.1007/978-3-031-78501-6_7

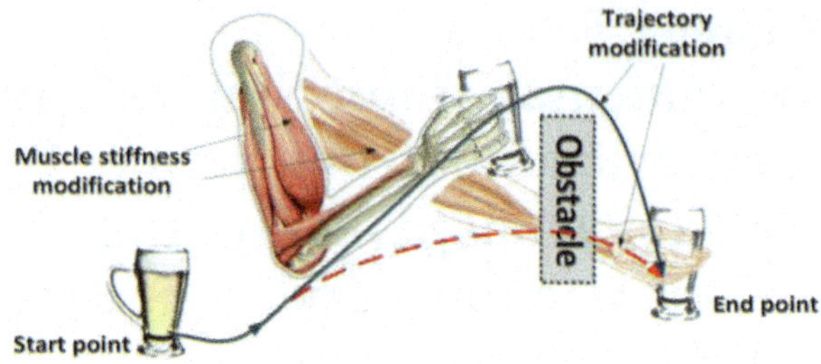

Fig. 7.1 Illustration of human dynamic reactions

the original function to enable the generalized to satisfy the special constraints [6, 11]. Hoffmann et al. [7] inspired by biological information and human behaviours to build an acceleration term calculated by the position of obstacle and velocity of robots to realize static and moving obstacle avoidance. Khansari-Zadeh and Billard [12] studied this issue by using dynamical systems to ensure the impenetrability of multiple convex-shaped obstacles. Gam et al. [13] proposed an adaptive DMP by adding a feedback term to the accelerating calculation of DMP for the instantaneous reactions. However, the feedback term is a simplified proportional control law with a constant gain. More recent research combined DMP with control methods for online path planning and control [14, 15], which is expected to develop to realize robot cooperation [19] or biped robot control [20]. For unknown obstacles, the above methods mostly choose to add an extra designed term to modify the DMP function. However, the solution for realizing obstacle avoidance is not unique and may be simultaneously affected by other constraints. So, the modified trajectory, after adding a force field term or a predesigned control diagram, may not be available in a way that humans would do.

To enable skills to be continually improved after adding new demonstrations, some researchers proposed integrating DMP and incremental learning method [1]. Kulić et al. [21] proposed an online, incremental learning of full-body motion primitives. Both motion primitives and primitive sequencing are learned and integrated in a framework to enable robots to complete humanoid motions. Lemme et al. also studied this issue and built a movement primitive library (MPL) by using self-supervised bootstrapping. The MPL is then used to compose and decompose complex trajectories by sequencing primitives. New movement primitive will be learned to extend the MPL to realize incremental learning [22]. However, these DMPs are placed as nodes to be connected to generate complex actions but are not updated along with newly added demonstrations. The recent research of Wang and Wu et al. proposed DMP plus method of inserting a bias term to the DMP implementation to let the original system learn a new given task accurately. Compared with other DMP-based incremental learning methods, the work of Wang and Wu et al. [8, 10] realized primitive skill-level learning, not merely the task level, that can enrich the usability of the existing primitives to create complex actions effectively. Additionally, some research presented methods for learning and generalizing primitive skills based on DMP to realize multiple stylistic skills [21–26] and skill adaptation [24] or interactive behaviour [25].

The above-mentioned studies are mostly about ordinary skill learning, and every subject is proposed with different usages and limitations, like DMP for obstacle avoidance cannot generalize humanoid style skills, and stylistic DMP can learn multiple types of skills but cannot address cases out of the plan and realize skill adaptation to the changes of the environment. Incremental learning-based DMP keeps learning skills from new demos but can't generate stylistic skills for the new cases in [23, 27]. To our knowledge, there is little research about the topic of stylistic incremental skill learning to acquire both ordinary and reactive skills to solve the task presented in Fig. 7.1. Additionally, the path points achieved by DMP can be utilized for multi-robot cooperation [19] or biped robot control [20] to realize obstacle avoidance and model-free motion tracking [15]. To address the above challenges, this chapter presents an incremental motor skill learning, generalization, and impedance control framework with the following contributions:

1. Proposing a hierarchical skill learning diagram to realize both ordinary skill and reactive skill learning, such that the learned results can cover the ordinary operations as well as cases out of the plan.
2. To realize stylistic reactive skill learning, the idea of the broad learning system (BLS) and DMP are integrated to enable the forcing function of DMP [a nonlinear term introduced in (7.3)] to keep updating by calculating stylistic factors, adding enhancement terms, and updating weight terms. The acquired reactive skills are expected to be combined with the ordinary skills, learned by the original DMP, to achieve an entire dynamic reaction.
3. To realize skill generalization with different conditions, an FLS-based stylistic skill generalization method is proposed. A fuzzifier is built to acquire a nonlinear reactive skill term with several fuzzy sets and rules about shapes, locations of the obstacle, human operational strategies, and muscle stiffness to create suitable reactions for varying environmental conditions. Then the skills can be generalized by changing not only start, end, and scaling as normal but also strategies, control impedances, and moving directions.
4. Based on the above contributions and our previous work of using the muscle stiffness for robot impedance control [17, 18], we proposed an adaptive impedance control based on the generalized outputs to realize the system stability with considerations of contact force estimation errors and robot uncertain dynamics.

7.2 Related Work

7.2.1 *Modelling of Robot*

Considering the following robot arm dynamics described by a Lagrangian formulation:

$$M\ddot{q} + C(\dot{q}, q)\dot{q} + G(q) + f(\dot{q}) = J(q)F_e + \tau, \qquad (7.1)$$

where q and $\dot{q}$ are n-dimensional vectors of robot joint position and velocity, M is a $n \times n$ symmetric positive-definite robotic inertia matrix, $C(\dot{q}, q)$ is a $n \times n$ matrix of Centripetal and Coriolis torque vector, $G(q)$ is the gradients of gravitational potential energy, $f(\dot{q})$ is the external friction vector, F_e is an environmental contact force, $J(q)$ is a Jacobian matrix and τ is the control torque to design. The system (7.1) has the following properties:

Property 7.1 *The matrix* $\dot{M} - 2C(\dot{q}, q)$ *is skew-symmetric.*

Property 7.2 *Jin et al. [28] set L as the smallest achievable sampling period in digital implementation, and the sampling rate is usually faster than 30 times. For the model in* (7.1), F_e *is considered as a hard nonlinearity satisfying* ${F_e}^{(t)} \neq {F_e}^{(t-L)}$ *as a discontinuous function, and* $f(\dot{q})$ *and* $J(q)$ *are continuous and soft nonlinearities estimated by time delay estimation (TDE) as* $f(\dot{q})^{(t)} \cong f(\dot{q})^{(t-L)}$ *and* $J(q)^{(t)} \cong J(q)^{(t-L)}$.

Property 7.3 *Changes of contact forces within the time interval L are* $\Delta {F_e}^{(t)} = {F_e}^{(t)} - {F_e}^{(t-L)}$ *and limited by*

$$\left|\Delta {F_e}^{(t)}\right| < \varepsilon, \tag{7.2}$$

where ε *is a constant and is estimated by measurements in actual.*

Property 7.4 *In any finite work space,* $J(q)$ *is nonsingular and bounded by* $|J(q)| \leq \alpha$.

7.2.2 Dynamic Movement Primitives

The DMP model for a task-space motion is presented as

$$\begin{cases} \tau\dot{v} = K(g - x) - Dv + (g - x_0)f(s) \\ \tau\dot{x} = v \end{cases}, \tag{7.3}$$

where $K, D > 0$ are damping and stiffness coefficients in [3], and $\tau > 0$ is a scaling parameter for adjusting the duration of the trajectory, and $f(s) = \theta^T \Psi(s)$ is a forcing function. where $\theta = [w_1, w_2, \dots, w_n]^T$, $\Psi(s) = [\psi_1, \psi_2, \dots, \psi_n]^T$ and ψ_i

$$\psi_i = \varphi_i(s)s \sum_{i=1}^{n} \varphi_i(s), \tag{7.4}$$

with Gaussian functions $\varphi_i(s) = \exp(-h_i(s - c_i)^2)$, where $c_i > 0$ and $h_i > 0$ are the centers and widths of radial basis functions. $f(s)$ depends on the phase variable s, which is calculated by a canonical system

$$\tau\dot{s} = -\gamma s, \quad \gamma > 0. \tag{7.5}$$

7.2.3 Broad Learning System

BLS is proposed based on the random vector functional-link neural network (RVFLNN) and inherits the merit of effectively eliminating the drawback of the long training process [29]. BLS is a flat network, where the original inputs are transferred as "mapped features" in the feature layer, and the structure is expanded in a wide sense in the "enhancement nodes" (as Fig. 7.2 shown). In recent years, BLS has been used for robotic control [9, 30], and micro aerial vehicle control. In Fig. 7.2, X is the input vector, and $Y \in R^{N \times C}$ is the output variable of the network, where N is the number of feature mappings and C is the dimension of the network's output. To explore hidden features of input data, the feature mappings are expressed as

$$Z_i = \phi_i \left(X W_{ei} + \beta_{ei} \right), \; i = 1, 2, \ldots, n, \tag{7.6}$$

where ϕ_i is a transformation function, and W_{ei} and β_{ei} are randomly sampled from the distribution density. Setting $Z^n = [Z_1, Z_2, \ldots, Z_n]$, and the jth enhancement term H_j of the functional link networks is generated by a linear function ξ_j, similar to ϕ_i, as

$$H_j = \xi_j \left(Z^n W_{hj} + \beta_{hj} \right), \, j = 1, 2, \ldots, m, \tag{7.7}$$

where W_{hj} and β_{hj} are randomly generated sampling data from the distribution density. Then the linked network is expanded by

$$\begin{aligned} Y &= \left[Z_1, \ldots, Z_n \left| \xi \left(Z^n W_{e1} + \beta_{e1} \right) \right., \ldots, \xi \left(Z^n W_{em} + \beta_{em} \right) \right] W \\ &= \left[Z_1, \ldots, Z_n \left| H_1, \ldots, H_m \right. \right] W \\ &= \left[Z^n \left| H^m \right. \right] W \\ &= A^{n+m} W \end{aligned} \,, \tag{7.8}$$

where $X^N = [X \,|x\,]$ represents the matrix X is extended in a row by x to achieve a new matrix X^N.

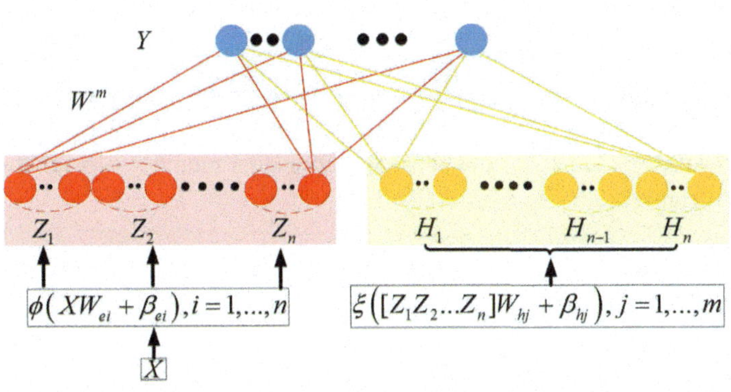

Fig. 7.2 Illustration of BLS

In (7.8), matrix W represents weights for the extended feature matrix $[F^n \,|H^m\,]$ and can be calculated by the pseudo-inverse calculation $W = [F^n \,|H^m\,]^+ Y$. After adding new feature terms H^m, W will be updated based on the calculation of the pattern matrix $[F^n \,|H^m\,]^+$. Setting $A^{n+1} = [A^n \,|a\,]$, where a is a new enhancement matrix, then the matrix W^{n+1} is updated by

$$W^{n+1} = \begin{bmatrix} W^n - db^T Y_n \\ b^T Y_n \end{bmatrix}, \tag{7.9}$$

where $d = (A^n)^+ a$, $b^T = \begin{cases} c^+ & if\ c \neq 0 \\ (1 + d^T d)^{-1} d^T (A^n)^+ & if\ c = 0 \end{cases}$, $c = a - A^n d$.

7.2.4 *Fuzzy Approximation*

Fuzzy approximation has been used for controlling robot [31], exoskeleton [32], pneumatic artificial muscle (PAM) [33], and mobile wheeled inverted pendulum [34]. Here, we consider an n_i inputs, a single-output fuzzy logic system with the product-inference rules, singleton fuzzifier, defuzzifier, and Gaussian membership function given by n_j fuzzy IF-THEN rules:

$$\begin{aligned} &R^{(j)} : \text{IF } x_1 \text{ is } A_1^j \text{ and } \ldots \text{ and } x_{n_i} \text{ is } A_{n_i}^j \text{ THEN } y \text{ is } B^j, \\ &[j = 1, \ldots, n_j] \end{aligned} \tag{7.10}$$

where $R^{(j)}$ denotes the jth rule, $\left(x_1, x_2, .., x_{n_j}\right)^T \in U$ and $y \in R$ are linguistic variables associated with FLS inputs and output respectively. A_i^j and B^j are fuzzy sets in U and R. Then the FLS performs a nonlinear mapping from U to R. We use the strategy of singleton defuzzification, and the output of product inference and center-average defuzzification, the FLS is

$$y(x) = \frac{\sum_{j=1}^m y_j \left(\prod_{i=1}^n \mu_{A_i^j}(x_i)\right)}{\sum_{j=1}^m \left(\prod_{i=1}^n \mu_{A_i^j}(x_i)\right)}, \tag{7.11}$$

where $x = [x_1, x_2, \ldots, x_m]^T$ and $\mu_{A_i^j}(x_i)$ represents membership functions of the linguistic variable x_i.

7.3 BLS-Based Skill Learning, Generalization and Control

In this section, we present a robot incremental skill learning and generalization, and the related impedance control framework, which consists of four stages coloured with different backgrounds in Fig. 7.3.

First, human operators create demonstrations in the known environment and dynamic reactions for the suddenly appeared obstacles (P1). For different obstacles, the demonstrating conditions e.g., human decisions, reactive motions and muscle stiffness are recorded. P2 is about skill learning. The ordinary skills and stylistic reactive skills are separately learned from normal actions and sudden reactions that are collectively called dynamic skills and introduced in detail in Sect. 7.3.1. As the reactive skills are determined by objective conditions, such as the size and location of obstacles as well as the decision of the actuator unit (human/robot), we describe these factors by fuzzy sets and use FLS to realize skill generalization in space and time with more interactions with the environment (P3, in Sect. 7.3.2). For a new task, trajectories and impedance factors are separately generalized and used for robot impedance control (P4, in Sect. 7.3.3).

7.3.1 BLS-Based Incremental Stylistic Skill Learning

First, we assume that the reactive motions are generated by new accidents. But they don't change the start and end points. Only the interval path points are changed between the start and destination. Then the new trajectory under the influence of obstacles with a new forcing term $f^N(s)$ is

$$\begin{cases} \tau\dot{v} = K(g - x) - Dv + (g - x_0)f^N(s) \\ \tau\dot{x} = v \end{cases}. \tag{7.12}$$

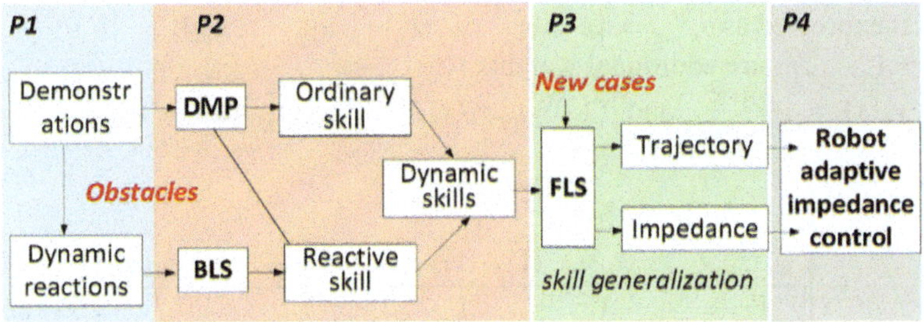

Fig. 7.3 Composition for dynamic stylistic skill learning and control framework

The term $f^N(s)$ is designed following the 'broaden' idea of BLS that part of Gaussian functions are chosen from the original $f(s)$, and some enhancement terms $\eta_j(s)$, $j = 1, 2, \ldots, m$ will be added into $f^N(s)$ to improve the nonlinear fitting effect to a new trajectory based on $\Phi(s) = [\varphi_1(s), \ldots, \varphi_n(s)]$, where $\varphi_i(s)$ is the Gaussian function shown in (7.4)

$$\eta_j(s) \equiv \zeta_j(\Phi(s)W_j + \beta_j), \tag{7.13}$$

where ζ_j is a transformation function similar to ξ in (7.7), and W_j and β_j are randomly generated weights.

The original $\varphi_j(s)$ and new added $\eta_j(s)$ will be normalized to generate new state terms $\lambda_k, k = 1, \ldots, m+n$:

$$\lambda_k = \frac{q_k(s)\varphi_k(s) + (1 - q_k(s))\eta_{k-n}(s)}{\sum_{i=1}^{n}\varphi_i(s) + \sum_{j=1}^{m}\eta_j(s)} s, \tag{7.14}$$

satisfying $\sum_{k=1}^{m+n}\lambda_k = s$, where $\begin{cases} q_k(s) = 1, & k \in [1, n] \\ q_k(s) = 0, & k \in [n+1, n+m] \end{cases}$.

Setting $\Gamma \sum_{i=1}^{n}\varphi_i(s) = \sum_{i=1}^{n}\varphi_i(s) + \sum_{j=1}^{m}\eta_j(s)$, we can get $\sum_{i=1}^{n}\varphi_i(s) + \sum_{j=1}^{m}\eta_j(s) = \frac{\Gamma}{\Gamma-1}\sum_{j=1}^{m}\eta_j(s)$. It is obvious that if terms $\varphi_i(s)$ and $\eta_j(s)$ are fixed, then Γ has a confirmed value. Then λ_k is expressed by ψ_i and $\eta_j(s)$ as:

$$\begin{aligned}\lambda_k &= \frac{q_k(s)\varphi_k(s)s}{\sum_{i=1}^{n}\varphi_i(s) + \sum_{j=1}^{m}\eta_j(s)} + \frac{(1 - q_k(s))\eta_{k-n}(s)s}{\sum_{i=1}^{n}\varphi_i(s) + \sum_{j=1}^{m}\eta_j(s)} \\ &= \frac{q_k(s)\varphi_k(s)s}{\Gamma\sum_{i=1}^{n}\varphi_i(s)} + \frac{(\Gamma-1)(1 - q_k(s))\eta_{k-n}(s)s}{\Gamma\sum_{j=1}^{m}\eta_j(s)} \\ &= \frac{q_k(s)\psi_k(s)}{\Gamma} + \frac{(\Gamma-1)(1 - q_k(s))\eta_{k-n}(s)s}{\Gamma\sum_{j=1}^{m}\eta_j(s)}\end{aligned}, \tag{7.15}$$

where $\eta_{k-n}(s) = 0$, if $k - n < 0$. Term λ_k corresponds with a new linear vector θ_{n+m}^N, which is expressed as $\theta_{n+m}^N = [w_1, w_2, \ldots, w_n | u_1, u_2, \ldots, u_m]^T = \left[\theta | \theta^U\right]^T$, where u_j, $j = 1, \ldots, m$ are additional weights.

Set $\gamma_j(s) = \frac{\eta_j(s)s}{\sum_{j=1}^{m}\eta_j(s)}$ and $\Upsilon(s) = \left[\gamma_1(s), \ldots, \gamma_m(s)\right]$, then

$$\begin{aligned}\Xi(s) &= \left[\lambda_1, \lambda_2, \ldots, \lambda_{n+m}\right]^T \\ &= \left[\frac{\psi_1(s)}{\Gamma}, \ldots, \frac{\psi_n(s)}{\Gamma} | \frac{(\Gamma-1)\gamma_1(s)}{\Gamma}, \ldots, \frac{(\Gamma-1)\gamma_m(s)}{\Gamma}\right]^T \\ &= \left[\frac{\Psi(s)}{\Gamma} | \left(1 - \frac{1}{\Gamma}\right)\Upsilon(s)\right]^T\end{aligned}. \tag{7.16}$$

Then the new forcing function $f^N(s)$ can be expressed as

$$\begin{aligned} f^N(s) &= \left(\theta_{n+m}^N\right)^T \Xi(s) \\ &= \left(\frac{1}{\Gamma}\right)\theta^T \Psi(s) + \left(1 - \frac{1}{\Gamma}\right)\left(\theta^U\right)^T \Upsilon(s) \\ &= \left(\frac{1}{\Gamma}\right) f(s) + \left(1 - \frac{1}{\Gamma}\right) f^U(s) \\ &= f(s) + \left(\frac{1}{\Gamma} - 1\right)\left(f(s) - f^U(s)\right) \end{aligned}, \tag{7.17}$$

where $f^U(s) = \left(\theta^U\right)^T \Upsilon(s)$.

Remark 7.1 The combination of BLS and DMP is realized by adding enhancement terms $\eta_j(s)$, $j = 1, \ldots, m$. Then the weights θ_{n+m}^N and state vector $\Xi(s)$ are extended from θ and $\Psi(s)$. By adding and normalizing more enhancement nodes, $f^N(s)$ will be improved to generate more complex trajectories.

Compared with (7.3), the error term

$$\begin{aligned} \Delta f(s) &= f^N(s) - f(s) \\ &= \left(\frac{1}{\Gamma} - 1\right)\left(f(s) - f^U(s)\right) \end{aligned}, \tag{7.18}$$

can be seen as a reactive skill adding to the ordinary one.

Term $\Delta f(s)$ can be expressed in a linear form as

$$\Delta f(s) = \theta_{n+m}^N \Xi^A(s), \tag{7.19}$$

where $\Xi^A(s) = \left[\left(\frac{1}{\Gamma} - 1\right)\Psi(s) | \left(1 - \frac{1}{\Gamma}\right)\Upsilon(s)\right]^T$.

Remark 7.2 The reactive skill term $\Delta f(s)$ in (7.19) is expressed in a similar form as $f(s)$. After adding a new term $\eta_j(s)$, θ_{n+m}^N and $\Xi^A(s)$ both will be extended and Γ will be changed as well. According to the theory of BLS, $\Delta f(s)$ can approach any nonlinear trajectory with the increase of enhancement nodes. Vector $\Xi^A(s)$ has two parts: one is $\left(\frac{1}{\Gamma} - 1\right)\Psi(s)$ representing changes of $\Psi(s)$ after adding new enhancement nodes and the other is $\left(1 - \frac{1}{\Gamma}\right)\Upsilon(s)$ to prolong the state vector.

In the DMP, the target value of $f(s)$ is calculated by

$$\min \sum_{j=1}^{n} \left\| f_j^{Tar}(s) - f(s) \right\|, \tag{7.20}$$

where

$$f_j^{Tar}(s) = \left(\tau \dot{v}_j - K\left(g - x_j\right) - D v_j\right) / \left(g - x_0\right), \tag{7.21}$$

and weights θ are learned by supervised learning algorithms such as the k-means clustering algorithm to minimize (7.20).

The target value of $f_j^N(s)^{Tar}$ is derived from (7.12) as

$$f_j^N(s)^{Tar} = \left(\tau \dot{v}_j^N - K\left(g - x_j^N\right) - Dv_j^N\right) / \left(g - x_0\right), \tag{7.22}$$

where x_j^N and v_j^N represent the position and velocity of the jth reactive trajectory. Then the desired value of $\Delta f(s)$ is acquired based on the calculation of $f(s)$ as

$$\begin{aligned}\Delta f_j(s)^{Tar} &= f_j^N(s)^{Tar} - f(s) \\ &= \frac{\left(\tau\left(\dot{v}_j^N - \dot{v}\right) - D\left(v_j^N - v\right) + K\left(x_j^N - x\right)\right)}{g - x_0},\end{aligned} \tag{7.23}$$

and then the reactive skill can be calculated by minimizing

$$\min \sum_{j=1}^{m} \left\| \Delta f_j^{Tar}(s) - \Delta f(s) \right\|. \tag{7.24}$$

Using (7.20) and (7.24), the ordinary skill and reactive skill will be acquired gradually. However, (7.24) is used to get only one group of θ_{n+m}^N to achieve one kind of reactive skill. To acquire multi-stylistic reactive skills, we propose the following method based on the previous equations. First, we assume that multiple reactive skills $\Delta f^k(s), k = 1, \ldots,$ K are derived from the common ordinary skill calculated by $f(s)$ and these reactive skills are learned based on the common state vector $\Xi^A(s)$ in (7.19) as

$$\Delta f^k(s) = \left(\theta_{n+m}^N\right)^k \Xi^A(s), \tag{7.25}$$

and the weights θ_{n+m}^N will be regressed into different values. Furthermore, we define a stylistic coefficient $\lambda_k, k = 1, 2, \ldots,$ K to normalize θ_{n+m}^N and (7.25) is expressed as

$$\Delta f^k(s) = \lambda_k \theta_{n+m}^N \Xi^A(s) = \lambda_k \Delta f^B(s). \tag{7.26}$$

In [21], Kulić et al. improved DMP by proposing stylistic-attractor landscapes instead of the forcing function. The weights for different styles are calculated based on the SVD calculation for target matrices. We also use SVD division for calculating stylistic factor λ_j based on the matrix $\Delta F(s)^{Tar} = \left[\Delta f_1(s)^{Tar}\right.$, $\left.\Delta f_2(s)^{Tar}, \ldots, \Delta f_K(s)^{Tar}\right]$ as

$$\Delta F(s)^{Tar} = U \Sigma V^T \approx \mathrm{M}\left(\Delta F^B(s)\right)^{Tar}, \tag{7.27}$$

and λ_j is the jth item of $\mathrm{M} = [\lambda_1, \lambda_2, \ldots, \lambda_K]$ and

$$\Delta F^B(s)^{Tar} = \left[\left(\Delta f_1^B(s)\right)^{Tar}, \left(\Delta f_2^B(s)\right)^{Tar}, \ldots, \left(\Delta f_K^B(s)\right)^{Tar}\right].$$

Then the weights θ_{n+m}^N are calculated by

$$\min \sum_{k=1}^{K} \left\| \left(\Delta f_k^B(s)\right)^{Tar} - \Delta f^B(s) \right\|. \tag{7.28}$$

The calculation procedure for the ordinary skill and reactive skills learning is realized by the following three steps:

1. **Step 6.1**: Using normal demonstrations and (7.3) and (7.5) to learn a general skill to approach the target function (7.20). The weights θ, state vector $\Psi(s)$, and forcing function $f(s)$ will be achieved after calculation.
2. **Step 6.2**: Human will make reactions for different situations, and the key influence factors leading to the reactions will be recorded along with reactions. By using (7.22) and (7.23), $\Delta f_k(s)^{Tar}$, $k = 1, \ldots,$ K are achieved and further transformed to $\left(\Delta f_k^B(s)\right)^{Tar}$ based on the SVD division.
3. **Step 6.3**: Using weights θ and state vector $\Psi(s)$ acquired in **step 1** and (7.13) to (7.19), we can initialize weights θ_{n+m}^N and $\Xi^A(s)$ first and then use (7.24) to optimize θ_{n+m}^N to achieve reactive skills. The weights θ_{n+m}^N will be prolonged after adding new enhancement nodes. For multi-stylistic reactive skill learning, we will choose (7.26) to replace (7.19) and use (7.28) for optimization.

7.3.2 FLS-Based Skill Generalization

DMP can realize skill generalization by changing the start, end and scaling factors, but the generalized results are similar to the demonstrated ones. Even if we can use the method in Sect. 7.3.1 to get some reactive skills to change the ordinary skill to some extent, the results still can't cover all the cases that have never arisen before. By using demonstrating conditions recorded in Step 6.2 and stylistic skills in Step 6.3, we build input fuzzy sets that have potential influences on the trajectories and muscle actions e.g., shapes of the obstacle, distance to the obstacle, and user's manipulating strategies, and IF-THEN rules to achieve reactive skills. For example, we set the distance from the start to the obstacle as t_1, the height of the obstacle as t_2, human moving direction as t_3, and the final reactive skills as output o. Then the n_j IF-THEN rules are described as

$$\begin{aligned}
&R^{(1)}\text{ :IF } t_1 \text{ is far and } t_2 \text{ is tall and } t_3 \text{ is left, THEN } o \text{ is } \Delta f^1(s);\\
&R^{(2)}\text{ :IF } t_1 \text{ is close and } t_2 \text{ is tall and } t_3 \text{ is right, THEN } o \text{ is } \Delta f^2(s);\\
&R^{(n_j)}\text{ :IF } t_1 \text{ is close and } t_2 \text{ is short and } t_3 \text{ is right, THEN } o \text{ is } \Delta f^{n_j}(s);
\end{aligned} \tag{7.29}$$

We select Gaussian fuzzy membership functions in (7.30), c_{ij} and σ_{ij} represent the center and width of the ith fuzzy set A_i^j

$$\mu_{A_i^j}(x_i) = \exp\left[-\frac{(x_i - c_{ij})^2}{\sigma_{ij}^2}\right]. \tag{7.30}$$

By using (7.11), the defuzzified reactive skill $\Delta f^F(s)$ is calculated by

$$\begin{aligned} \Delta f^F(s) &= \frac{\sum_{j=1}^{n_j} \Delta f^j(s) \left(\prod_{i=1}^{n} \mu_{A_i^j}(x_i)\right)}{\sum_{j=1}^{n_j} \left(\prod_{i=1}^{n} \mu_{A_i^j}(x_i)\right)} \\ &= \frac{\sum_{j=1}^{n_j} \lambda_j \theta_{n+m(j)}^N \Xi^A(s) \left(\prod_{i=1}^{n} \mu_{A_i^j}(x_i)\right)}{\sum_{j=1}^{n_j} \left(\prod_{i=1}^{n} \mu_{A_i^j}(x_i)\right)} \end{aligned}. \tag{7.31}$$

Remark 7.3 It is proven that FLS can uniformly approximate any given continuous function over a compact set to any degree of accuracy. (7.31) can also be expressed as

$$\Delta f^F(s) = \frac{\sum_{j=1}^{m} \lambda_j \theta_{n+m(j)}^N \left(\prod_{i=1}^{n} \mu_{A_i^j}(x_i)\right)}{\sum_{j=1}^{m} \left(\prod_{i=1}^{n} \mu_{A_i^j}(x_i)\right)} \Xi^A(s), \tag{7.32}$$

which can be seen as a nonlinear composition of $\theta_{n+m(j)}^N$ in (7.19) based on the same vector $\Xi^A(s)$ to generalize modified stylistic reaction skills. Thus, the reactive skills (7.31) can approximate any reactions within the designed ranges.

The main problem that occurred in the fuzzy skill generalization process was the scope changes of the fuzzy sets. Because the fuzzy rules generated in the demonstrations may not fit new tasks, such as for a 10 cm trajectory, 8 cm is a long distance, but for a 100 cm trajectory, it is a short distance. Thus the fuzzy sets and IF-THEN rules should be changed along with the ordinary skills generalization. Define the start and end in a new task as G and X_0, sampling time interval as T and trajectory position as X, using the learned function $f(s)$ in (7.3), the skill generalizations satisfy

$$\begin{cases} T\dot{V} = K(G - X) - DV + (G - X_0) f(s) \\ T\dot{X} = \dot{V} \end{cases}. \tag{7.33}$$

Factors c_{ij} and σ_{ij} in (7.30) are reset as C_{ij} and Δ_{ij} following the new generated trajectory point X_i as

$$\mu_{A_i^j}(X_i) = \exp\left[-\frac{(X_i - C_{ij})^2}{\Delta_{ij}^2}\right], X_i \in \left[\underline{X}_i, \bar{X}_i\right]. \tag{7.34}$$

As c_{ij} and σ_{ij} in (7.30) are chosen based on human experience, the factors C_{ij} and Δ_{ij} in [1] also rely on humans' selections. They can keep the same values as in (7.30), enlarge or decrease the values several times for c_{ij} and σ_{ij} to enable fuzzy sets to cover the scope of the whole task space. Then the defuzzified reactive skills in (7.31) are revised as

$$\Delta f^F(s) = \frac{\sum_{j=1}^{m} \lambda_j \theta_{n+m(j)}^{N} \Xi^A(s) \left(\prod_{i=1}^{n} \mu_{A_i^j}(X_i) \right)}{\sum_{j=1}^{m} \left(\prod_{i=1}^{n} \mu_{A_i^j}(X_i) \right)}. \tag{7.35}$$

Then the final generalized dynamic skill is expressed as

$$\begin{cases} T\dot{V} = K(G - X) - DV + (G - X_0)\left(f(s) + \Delta f^F(s)\right) \\ T\dot{X} = \dot{V} \end{cases}. \tag{7.36}$$

Remark 7.4 In our previous work [17, 18], humanlike variable impedance skills are abstracted from EMG signals for robot control. However, the values of impedance cannot be enlarged or reduced freely in a similar way to the trajectories due to the physiological limitations. Some IF conditions are determined by exact space metrics like the height or position of the obstacle and can be quantized, while others depend on demonstrators' habits and minds, like turning directions or nervous tensions, which are expected to be held as the same IF conditions as in demonstrations.

7.3.3 Variable Impedance Control

Some recent studies of EMG signal recognition are used for exoskeleton [35] and myoelectric hand control [36]. To enable robots to realize the human-like control effect, some studies about skills learning used EMG signals for updating factors of DMP and realizing DMP-based variable impedance control. In our previous research [17, 18], the impedance is generalized and used for enduring heavier objects in the vertical direction. While in [37], the "variable impedance behaviour in robots is the active safety due to the soft 'giving in' for both robots and the environment". As the experimental results shown in the next section, when the manipulator end approaches an obstacle, the stiffness of the human handling arm will increase to avoid conflict. Here, we set a controlling factor K_{imp} that has a linear relation to muscle stiffness X^{stf} as

$$K_{imp} = k_s X^{stf}, \tag{7.37}$$

where k_s is a constant factor. Meanwhile, some manipulators don't provide force sensors and $f(\dot{q})$ in (7.1) is hard to model accurately. For dynamics errors and unknown force information, we used neural networks (NN) for force sensorless admittance control previously [38, 39]. Following Property 7.2, we use time-delayed estimation (TDE) for compensating these error terms in this chapter. Define q^d as the desired value of q, $e = q^d - q$, $\dot{e} = \dot{q}^d - \dot{q}$, and $r = \dot{e} + ke$, where k is a constant and $\tau_e = f(\dot{q}) - JF_e$, then the controller is designed as

$$\tau = M\left(\ddot{q}^d + k\dot{e}\right) + C\left(\dot{q}^d + ke\right) + G(q) - \widehat{\tau_e}, \tag{7.38}$$

where J represents a Jacobian matrix, and $\widehat{\tau_e}$ is a control term that is deduced by the delayed signals as

$$\begin{aligned} \widehat{\tau_e} &= {\tau_e}^{(t-L)} + K_{imp}r \\ &= M\ddot{q}^{(t-L)} + C\dot{q}^{(t-L)} + G(q)^{(t-L)} - \tau^{(t-L)} + K_{imp}r' \end{aligned} \tag{7.39}$$

where K_{imp} is an impedance-related positive factor that is expressed as

$$K_{imp} = \max\left(k_s X^{stf}, sign(r)\Gamma\varepsilon r^{-1}\right), \tag{7.40}$$

where $\Gamma >> \alpha$ is a large enough factor. (7.37) is improved by considering the estimation errors to achieve (7.40).

Due to the dynamics modelling and measuring errors, selecting an unsuitable constant k_s may cause system instability. The impedance factor in (7.40) is designed to realize human-like muscle stiffness changes as well as to ensure system stability, which is proved by the following Lyapunov function:

$$V = r^T M r. \tag{7.41}$$

Taking (7.38) into (7.1) and using Step 6.3, we have

$$M\dot{r} + \left(C + K_{imp}\right) r = \tau_e - {\tau_e}^{(t-L)} = J\left(F_e - {F_e}^{(t-L)}\right). \tag{7.42}$$

Taking (7.42) into the time derivative of V, we have

$$\begin{aligned} \dot{V} &= 2\dot{r}^T M r + r^T \dot{M} r \\ &= 2J\left(F_e - {F_e}^{(t-L)}\right) r + r^T \dot{M} r - 2r^T C r - 2r^T K_{imp} r. \\ &= 2J\Delta F_e r - 2r^T K_{imp} r \le 2\alpha\varepsilon r - 2r^T K_{imp} r \end{aligned} \tag{7.43}$$

Using (7.40), we have $\frac{sign(r)K_{imp}r}{\Gamma} \ge \varepsilon$, then (7.43) can be simplified as

$$\dot{V} \le 2\left(\frac{\alpha}{\Gamma} - 1\right) r^T K_{imp} r < 0. \tag{7.44}$$

7.4 Experiments

The experimental platform in Fig. 7.4. consists of a Touch Omni joystick, a 3D-print finger handle, a 21 cm × 13 cm operational board, a 3D-print obstacle, and a MYO armband for measuring EMG signals of the operator's limb. The tool is fixed at the tip of the Touch joystick and can be replaced by a gripper driven by a servo motor. The height of the obstacle can be modified by adding or reducing several 2 cm × 2 cm × 2 cm cubes. Using the 3D-print handle, the operator can put a finger into a fixing hole and drag the end of the joystick to move from the start (23,0) to the end (20,90), and the Touch joystick will record the joints and end positions. At the same time, human muscle EMG signals are also recorded. To acquire different demonstrations from start to end, we change operational situations (presented in Table 7.1) by adding obstacles with different heights and placing them at different positions. The demonstrator will choose three directions (right, middle, and left) to cross obstacles, and the various reactions will be recorded. Figure 7.5 shows a demonstration with a high obstacle placed at the long-distance position (13,5).

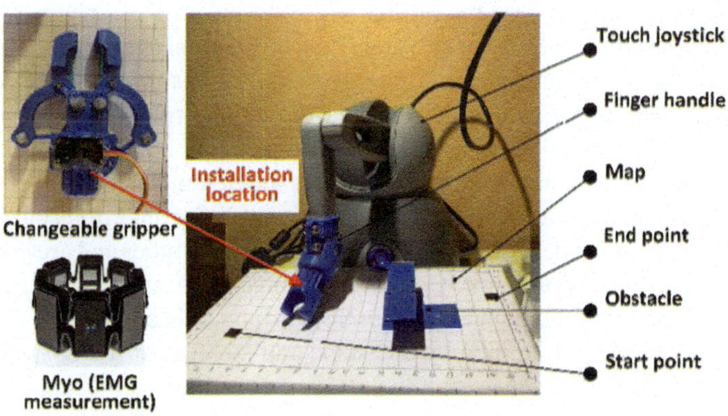

Fig. 7.4 Experimental platform

Table 7.1 Factors and experimental conditions

Factors	Experimental conditions
Distance to obstacle	Short distance
Height of obstacle	Tall (8 cm)
	Low (2 cm)
Direction	Left
	Forward
	Right

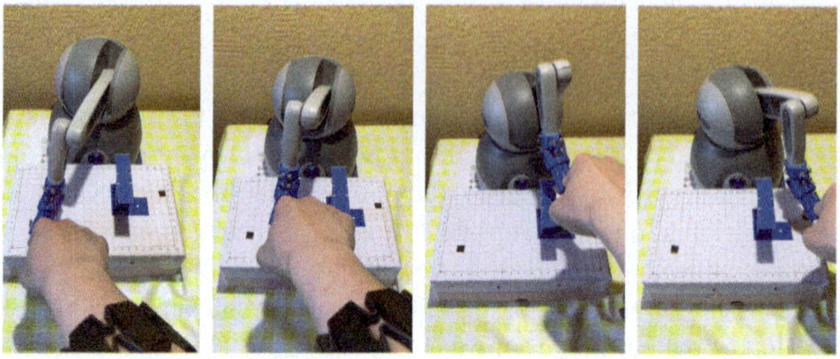

(a) (b) (c) (d)

Fig. 7.5 Process of demonstrations (high obstacle placed at the far location). **a** Start stage. **b** Move away from the start point. **c** Cross over the obstacle. **d** Reach the target point

7.4.1 DMP/BLS-Based Skill Learning

Following the experimental procedure (Step 6.1 to Step 6.3) shown in Sect. 7.3.1, we firstly demonstrate and record the ordinary actions and stiffness changes (grey dash lines in Fig. 7.6a and b. By using (7.3) and (7.5), the general skills are abstracted from the demonstrations (red lines in Fig. 7.6a and b as the basis and benchmark case for the reactive skills learning. The trajectories measured in the special conditions in Table 7.1 are presented from Fig. 7.6c–f. Three coloured line bundles show trajectories with different moving directions. We can get a rough conclusion that for the lower obstacle, the trajectories are easy to distinguish in different directions but for a higher obstacle, the forward and right motions override with each other and are hard to divide due to the mechanical design of 3D tool and limited working space of the joystick.

Figure 7.7 shows impedance changes for different reactions. We find that the impedance is mainly determined by the shape and location of the obstacle and does not have much correspondence with the moving directions on the same level. Thus, we present impedance changes in four figures divided by different obstacle heights and locations. The comparison of (a) and (b) shows that the impedance increases with the height of the obstacle by changing from 1 block to 3 blocks. The comparison of (a) and (c), and (b) and (d) with the height but different object locations show the tiny changes affected by the increasing speed of impedance such that when the obstacle is moved toward the start, operators will enlarge impedance ahead for crossing over obstacles. But the final reaching values are similar.

Following Step 6.2 and Step 6.3 in Sect. 7.3.1, the reactive motions are derived from the ordinary one in Fig. 7.6 by adding enhancement nodes and extending members of weights and the common state vector $\Xi^A(s)$ in (7.25). Figure 7.8a–c present the learning process after adding 1, 2, and 5 enhancement terms for the task with a lower obstacle and short-distance location. With the increase of enhancement nodes and updating of weights, the trajectories are expanded in three directions from the

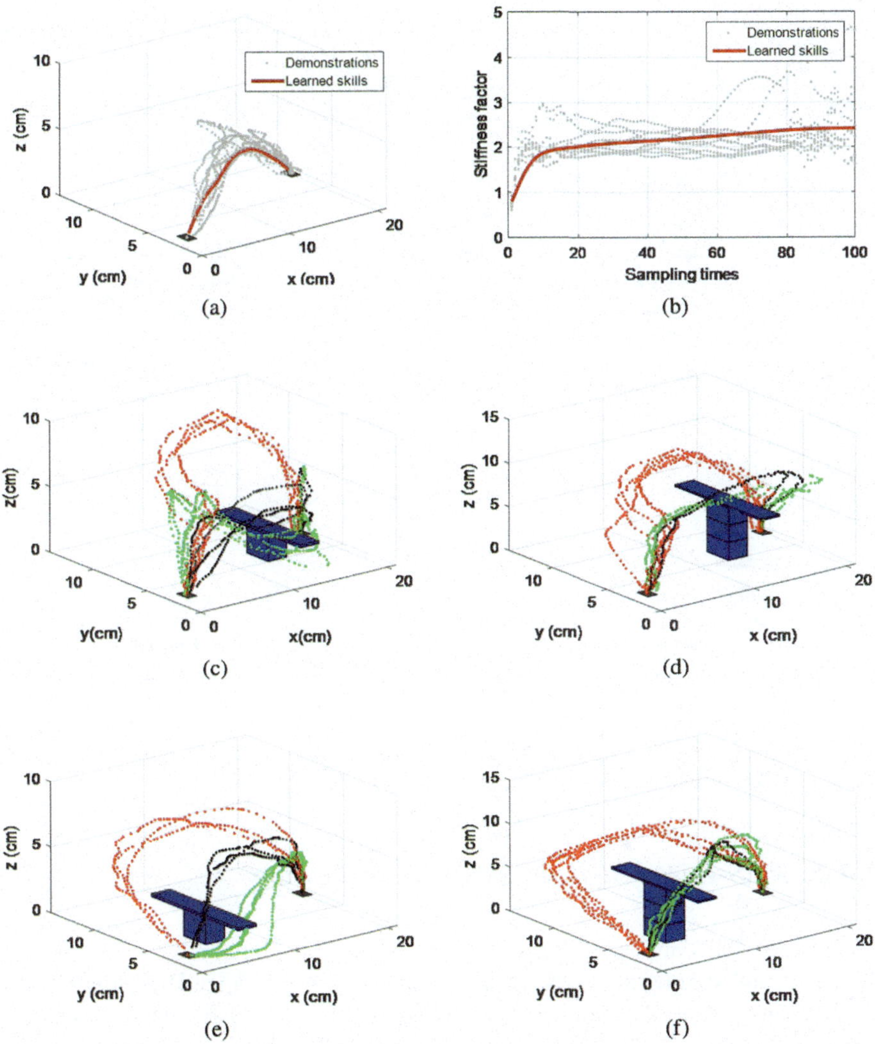

Fig. 7.6 Skill demonstrations and ordinary skill learning. **a** Ordinary skill learning of trajectories. **b** Ordinary skill learning of muscle stiffness. **c** Trajectories with a low obstacle at the far location. **d** Trajectories with a high obstacle at the far location. **e** Trajectories with a low obstacle at the close location. **f** Trajectories with a high obstacle at the close location

ordinary to approach the classified demonstrated reactions gradually. Figure 7.8d shows the final learned results as well as demonstrated reactions. The learned reactive skills can represent the classified trajectory groups with minor errors.

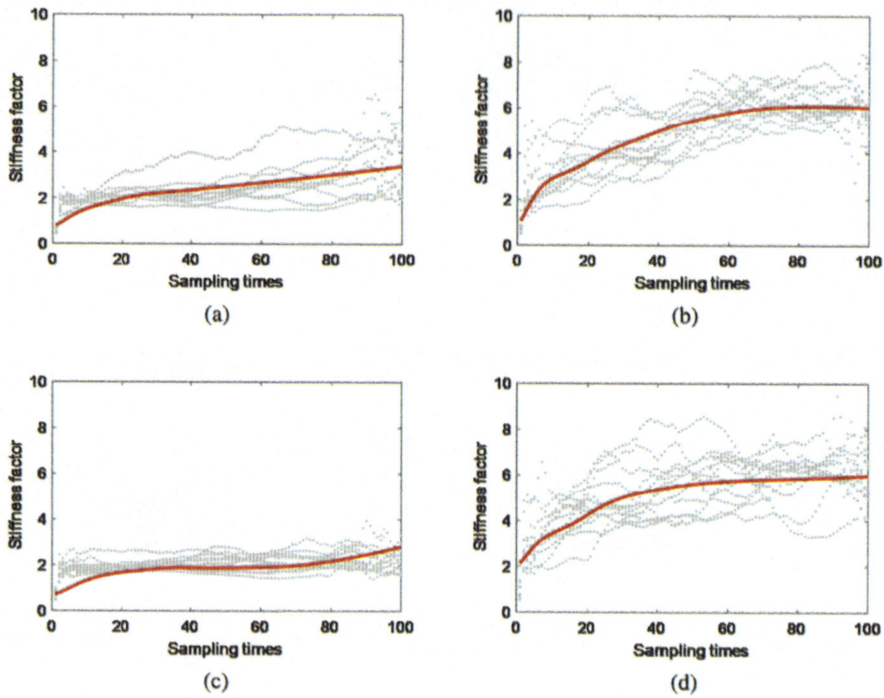

Fig. 7.7 Stiffness changes for different cases. **a** Results with a low obstacle at the far location. **b** Skill results in a high obstacle at the far location. **c** Results with a low obstacle at the close location. **d** Skill results in a high obstacle at the close location

7.4.2 FLS-Based Stylistic Skill Generalization

The learned skills are generalized through several steps. First, after knowing a new start, end, and scaling factor, the ordinary trajectory is generalized by using DMP. In Fig. 7.9a, we change the start to (46,0) and the endpoint to (18,40). The new ordinary trajectory is coloured in blue, compared with the old one coloured in red. Second, we place a medium-height obstacle (4 cm) at (10,50). Because the new trajectory will be generated on the same map, the IF-THEN rules are chosen as same as in the demonstrations. Considering the influence of the environment, we generalize the skill in three typical directions, and the results are shown in Fig. 7.9b. Compared to the cases with lower and higher obstacles, the overlaps of forward and right trajectories stand at the medium level. Additionally, the moving directions are determined by the operator's strategies not by angles, thus we choose dimensionless standards to express the transitional state between the right and left trajectories, and the centres of fuzzy sets are 0, 0.8, and 1. Finally, using (7.35), the movements with different directions are generalized as grey lines in Fig. 7.9c, and we choose a left-forward direction as the final trajectory as the red line shown in Fig. 7.9c. As mentioned

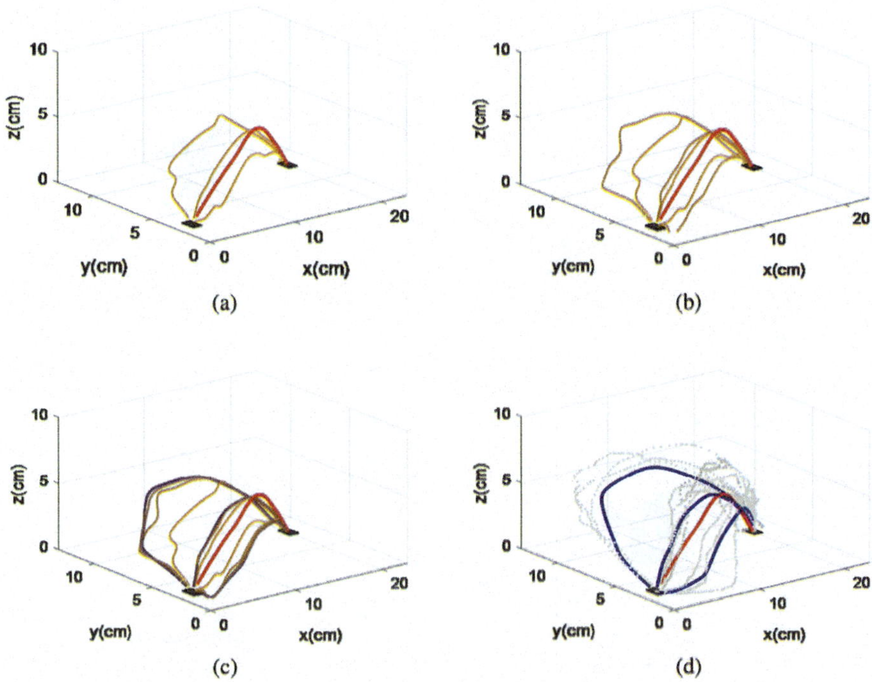

Fig. 7.8 Reactive trajectory learning process. **a** Results of adding one group of enhancement nodes. **b** Results of adding two groups of enhancement nodes. **c** Results of adding five groups of enhancement nodes. **d** Final learned results and classified demonstrations

in Remark 7.4, the stiffness has a close relationship with the obstacle's height and positions, thus we choose the average value of stiffness in lower and higher cases as start and end for impedance generalization. The final stiffness generalization results are shown in Fig. 7.9d.

7.4.3 Adaptive Impedance Control

To enable the joystick to work as an actuator, we choose the kinematic and dynamics estimation method introduced in [40]. The factors in (7.40) are set as $k_s = 10, \varepsilon = 1,$ $\Gamma = 100$, and the experiment process from closing gripper to picking and placing the object to the destination are shown in Fig. 7.10, and the upper joint torques varying with time are shown in Fig. 7.11a–c. The adaptive impedance ensures system stability and tracking performance of the gripper to the trajectory generalized by FLS Fig. 7.11d.

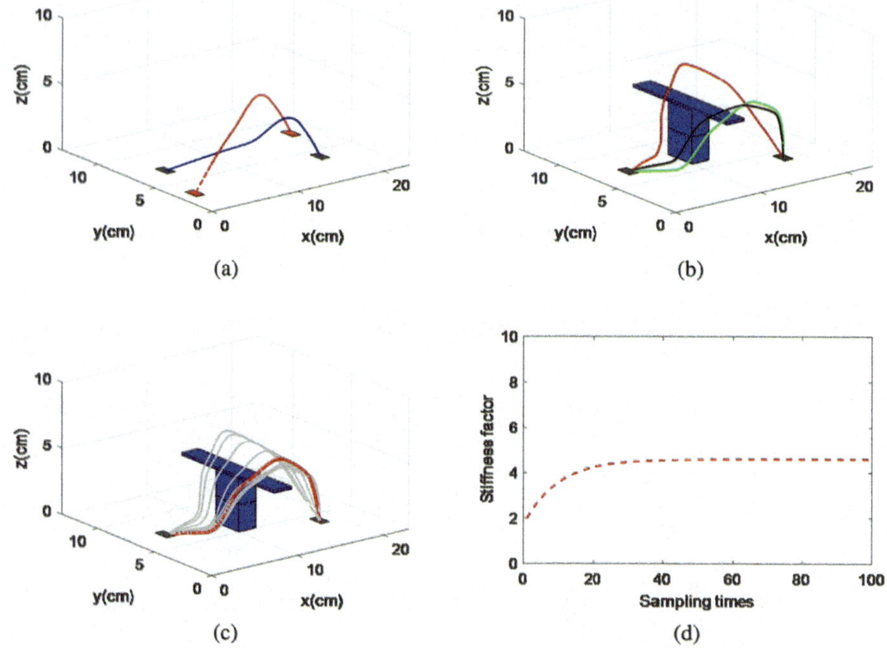

Fig. 7.9 FLS-based skill generalization. **a** Ordinary skill generalization with new start and end points. **b** Skill generalization with environmental factors (height and location of the obstacle). **c** Skill generalization with environmental factors and operational strategies (moving directions). **d** Generalization of muscle stiffness

The main difference between the proposed method to the previous methods is dividing skills into an ordinary skill and a reactive skill. The ordinary skill is learned ahead, and reactive skills are incrementally learned based on the results of the ordinary skill. Thus, both two kinds of skills are kept and can be generalized separately and cooperatively. The FLS method extends dimensions for skill generalization, not only in space and time but also with objective conditions such as environmental influences, human strategies, and muscle stiffness by building multiple fuzzy sets.

7.5 Conclusion

In this chapter, we proposed an incremental skill learning and generalization method based on dynamic movement primitive, fuzzy logic, and broad learning system and used the generalized stiffness for robotic impedance control. Different from previous skill learning methods from demonstrations, the method learns from human ordinary actions and reactions to sudden incidents comprehensively. Thus, the pro-

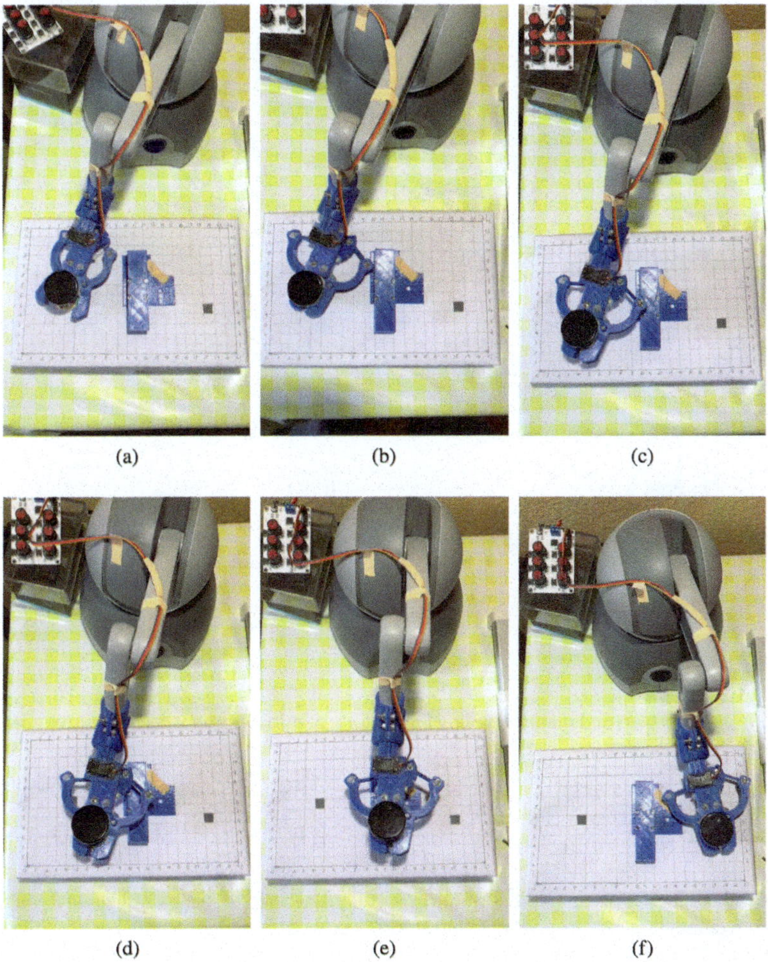

Fig. 7.10 Manipulation process by using a gripper. **a** Grasp an object. **b–f** Object moving process

posed method can realize skill generalization not only in space and time but also considering environmental changes, muscle stiffness, and decisions. The technical effectiveness is verified through an experiment in which an object is picked by a desk robot (modified joystick) to cross over an unknown obstacle to finally be placed at the target. A brief discussion shows that the proposed method can realize life-long learning and endure more complex operational tasks.

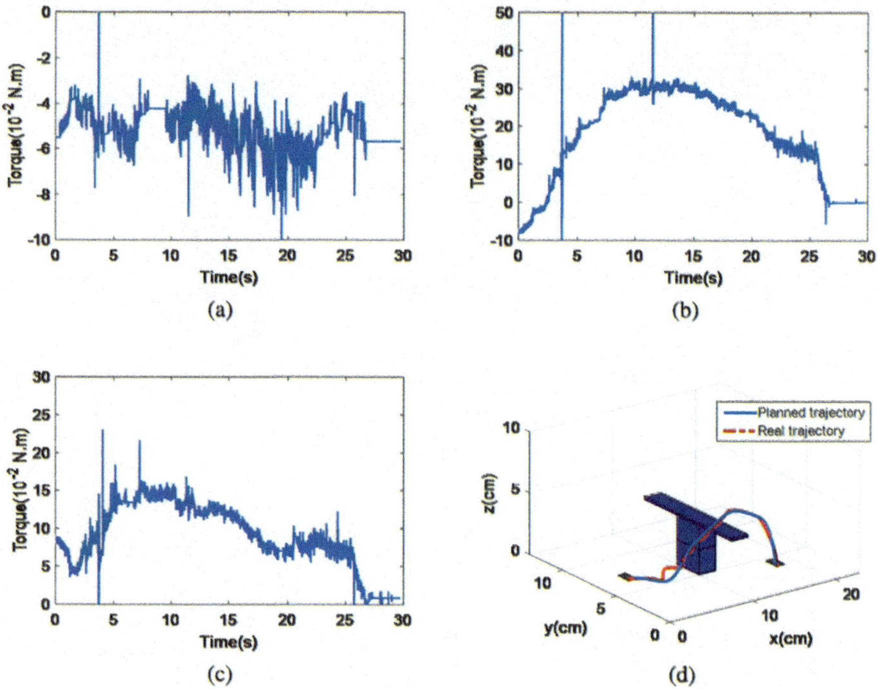

Fig. 7.11 Joint torques and planned and real trajectory. **a–c** Joint torque. **d** Trajectories comparison

References

1. Ravichandar H, Polydoros AS, Chernova S, Billard A (2020) Recent advances in robot learning from demonstration. Annu Rev Control Rob Autonom Syst 3:297–330
2. Schaal S (2006) Dynamic movement primitives—a framework for motor control in humans and humanoid robotics. In: Adaptive motion of animals and machines. Springer, Tokyo
3. Saveriano M, Abu-Dakka FJ, Kramberger A, Peternel L (2021) Dynamic movement primitives in robotics: a tutorial survey. Int J Rob Res 42(13):1133–1184
4. Li Z, Zhao T, Chen F, Hu Y, Su CY, Fukuda T (2017) Reinforcement learning of manipulation and grasping using dynamical movement primitives for a humanoid like mobile manipulator. IEEE/ASME Trans Mechatron 23(1):121–131
5. Zhao T, Deng M, Li Z et al (2018) Cooperative manipulation for a mobile dual-arm robot using sequences of dynamic movement primitives. IEEE Trans Cogn Dev Syst 12(1):18–29
6. Park DH, Hoffmann H, Pastor P et al (2008) Movement reproduction and obstacle avoidance with dynamic movement primitives and potential fields. Paper presented at the humanoids 2008—8th IEEE-RAS international conference on humanoid robots, Daejeon, Korea (South), 01–03 December 2008
7. Hoffmann H, Pastor P, Park DH et al (2009) Biologically-inspired dynamical systems for movement generation: Automatic real-time goal adaptation and obstacle avoidance. Paper presented at the 2009 IEEE international conference on robotics and automation, Kobe, Japan, 12–17 May 2009
8. Wang R, Wu Y, Chan WL et al (2016) Dynamic movement primitives plus: For enhanced reproduction quality and efficient trajectory modification using truncated kernels and local

biases. Paper presented at the 2016 IEEE/RSJ international conference on intelligent robots and systems (IROS), Daejeon, 09–14 October 2016
9. Huang H, Yang C, Chen CLP (2020) Optimal robot-environment interaction under broad fuzzy neural adaptive control. IEEE Trans Cybern 51(7):3824–3835
10. Wu Y, Wang R, D'Haro LF et al (2018) Multi-modal robot apprenticeship: imitation learning using linearly decayed dmp+ in a human-robot dialogue system. Paper presented at the 2018 IEEE/RSJ international conference on intelligent robots and systems (IROS), Madrid, Spain, 01–05 October 2018
11. Tan H, Erdemir E, Kawamura K et al (2011) A potential field method-based extension of the dynamic movement primitive algorithm for imitation learning with obstacle avoidance. Paper presented at the 2011 IEEE international conference on mechatronics and automation, Beijing, China, 07–10 August 2011
12. Khansari-Zadeh SM, Billard A (2012) A dynamical system approach to realtime obstacle avoidance. Autonom Rob 32:433–454
13. Gams A, Petrič T, Do M et al (2016) Adaptation and coaching of periodic motion primitives through physical and visual interaction. Rob Autonom Syst 75:340–351
14. Rai A, Sutanto G, Schaal S et al (2017) Learning feedback terms for reactive planning and control. Paper presented at the 2017 IEEE international conference on robotics and automation (ICRA), Singapore, 29 May 2017–03 June 2017
15. Khan AT, Li S, Li Z (2021) Obstacle avoidance and model-free tracking control for home automation using bio-inspired approach. Adv Control Appl 4(1):e63
16. Lu Z, Wang N, Yang C (2021) A constrained DMP framework for robot skills learning and generalization from human demonstrations. IEEE/ASME Trans Mechatron 26(6):3265–3275
17. Yang C, Zeng C, Fang C et al (2018) A DMP-based framework for robot learning and generalization of humanlike variable impedance skills. IEEE/ASME Trans Mechatron 23(3):1193–1203
18. Yang C, Zeng C, Liang P et al (2018) Interface design of a physical human-robot interaction system for human impedance adaptive skill transfer. IEEE Trans Autom Sci Eng 15(1):329–340
19. Khan AT, Li S, Cao X (2021) Control framework for cooperative robots in smart home using bio-inspired neural network. Measurement 167:108253
20. Khan AT, Li S, Zhou X (2021) Trajectory optimization of 5-link biped robot using beetle antennae search. IEEE Trans Circ Syst II Exp Briefs 68(10):3276–3280
21. Kulic D, Ott C, Lee D et al (2012) Incremental learning of full body motion primitives and their sequencing through human motion observation. Int J Rob Res 31(3):330–345
22. Lemme A, Reinhart RF, Steil JJ (2014) Self-supervised bootstrapping of a movement primitive library from complex trajectories. Paper presented at the 2014 IEEE-RAS international conference on humanoid robots, Madrid, Spain, 18–20 November 2014
23. Matsubara T, Hyon SH, Morimoto J (2011) Learning parametric dynamic movement primitives from multiple demonstrations. Neural Netw 24(5):493–500
24. Zhao Y, Xiong R, Fang L et al (2014) Generating a style-adaptive trajectory from multiple demonstrations. Int J Adv Rob Syst 11(7):103
25. Young JE, Igarashi T, Sharlin E et al (2014) Design and evaluation techniques for authoring interactive and stylistic behaviors. ACM Trans Interact Intell Syst (TiiS) 3(4):1–36
26. LaViers A, Egerstedt M (2014) Style-based abstractions for human motion classification. Paper presented at the 2014 ACM/IEEE international conference on cyber-physical systems (ICCPS), Berlin, Germany, 14–17 April 2014
27. Matsubara T, Hyon SH, and Morimoto J (2010) Learning stylistic dynamic movement primitives from multiple demonstrations. Paper presented at the 2010 IEEE/RSJ international conference on intelligent robots and systems, Taipei, Taiwan, 18–22 October 2010
28. Jin M, Kang SH, Chang PH (2008) Robust compliant motion control of robot with nonlinear friction using time-delay estimation. IEEE Trans Indus Electron 55(1):258–269
29. Liu Z et al (2017) Broad learning system: feature extraction based on K-means clustering algorithm. Paper presented at the 2017 4th international conference on information, cybernetics and computational social systems (ICCSS), Dalian, China, 24–26 July 2017

30. Huang H, Zhang T, Yang C, Chen CP (2019) Motor learning and generalization using broad learning adaptive neural control. IEEE Trans Indus Electron 67(10):8608–8617
31. Yang C, Jiang Y, Na J et al (2018) Finite-time convergence adaptive fuzzy control for dual-arm robot with unknown kinematics and dynamics. IEEE Trans Fuzzy Syst 27(3):574–588
32. Li Z, Su CY, Wang L et al (2015) Nonlinear disturbance observer-based control design for a robotic exoskeleton incorporating fuzzy approximation. IEEE Trans Indus Electron 62(9):5763–5775
33. Chen C, Huang J, Wu D et al (2021) Interval type-2 fuzzy disturbance observer-based TS fuzzy control for a pneumatic flexible joint. IEEE Trans Indus Electron 69(6):5962–5972
34. Huang J, Ri M, Wu D et al (2017) Interval type-2 fuzzy logic modeling and control of a mobile two-wheeled inverted pendulum. IEEE Trans Fuzzy Syst 26(4):2030–2038
35. Huang J, Huo W, Xu W et al (2015) Control of upper-limb power-assist exoskeleton using a human-robot interface based on motion intention recognition. IEEE Trans Autom Sci Eng 12(4):1257–1270
36. Xing K, Yang P, Huang J et al (2014) A real-time EMG pattern recognition method for virtual myoelectric hand control. Neurocomputing 136:345–355
37. Hu Y, Wu X, Geng P et al (2019) Evolution strategies learning with variable impedance control for grasping under uncertainty. IEEE Trans Indus Electron 66(10):7788–7799
38. Yang C, Peng G, Li Y et al (2018) Neural networks enhanced adaptive admittance control of optimized robot-environment interaction. IEEE Trans Cybern 49(7):2568–2579
39. Yang C, Peng G, Cheng L et al (2019) Force sensorless admittance control for teleoperation of uncertain robot manipulator using neural networks. IEEE Trans Syst Man Cybern Syst 51(5):3282–3292
40. Sansanayuth T, Nilkhamhang I, Tungpimolrat K (2012) Teleoperation with inverse dynamics control for phantom omni haptic device. Paper present at the 2012 proceedings of SICE annual conference (SICE), Akita, Japan, 20–23 August 2012

Chapter 8
Broad Fuzzy Neural Adaptive Impedance Control for Optimal Robot-Environment Interaction

8.1 Introduction

During the past few years, increasing robotic manipulator techniques have been employed in industrial applications and our daily lives. One of the most difficult problems in robotic research is how to achieve a human-like working manner in its workspace, such as the distinctive capability of skills learning and friendly desirable interaction control [1–3]. Such abilities are of great importance to enable robots to respond more intelligently and be able to fulfil more complex, dexterous and versatile tasks in various fields such as industrial applications and service areas.

In our daily lives, humans can accomplish complex tasks with great ease by inspiring and generalising a set of fundamental skills from the past. Notably, it would be outstanding for the robotic control systems to benefit from such brilliant learning capability. In the last decades, many researchers have paid massive efforts to investigate the robotic motor control learning problems to provide reasonable decision-making and advanced control methods, such as sliding mode control [4, 5], adaptive NN contorl [6, 7] and robust control [8, 9], etc. In general, it is well known that the robot dynamic is hard to be modelled appropriately with the increasing complexity of the manipulator's structure. Due to the excellent approximation capability of NN [10] and fuzzy system [11, 12], they have been extensively utilized for adaptive control systems to estimate the unknown dynamic of robots at any specific accuracy. Particularly, a notable method which incorporates the fuzzy logic to adaptive NN control has been widely adopted to address the problems of lacking automatic learning capabilities in fuzzy system [13–15]. By considering the uniform boundedness and state constraints, He et al. [16] proposed an adaptive fuzzy neural network (FNN) control to address the interaction problem of the robot. In [17], a developed FNN control structure is employed to approximate the uncertain dynamic model of a dual-arm robot. Despite much progress achieved in model approximation, several other methodologies have been proposed to ensure the control performance, such as input saturation [18] which can tackle the control problem related to input constraint, finite-time convergence control [19].

C. Yang et al., *Advanced Teleoperation and Robot Learning for Dexterous Manipulation*, Springer Tracts in Advanced Robotics 160,
https://doi.org/10.1007/978-3-031-78501-6_8

Nevertheless, a key problem in the aforementioned control schemes which is tough to address is that the neural nodes for NN or the logic rules for the fuzzy system are specified beforehand in terms of the expert experiences. Once the desired motion trajectory or working environment is changed, the original controller, unsurprisingly, may result in poor control tracking performance and even irreversible consequences. In such a case, the FNN controller, which lacks generation ability from a neural biological attribute point of view, needs to be redesigned to deal with the new tasks. Recently, a novel framework named broad learning system [20], which is developed from random vector functional-link neural networks (RVFLNN) [21], has been proposed and extensively utilized for pattern recognition and classification [22–24]. By taking advantage of RVFLNN, BLS can offer acceptable generalization and efficient expansion performance by increasing the neural nodes dynamically. Herein, this inspires us to bring a forward solution which incorporates BLS to improve the generalizing ability of FNN.

As mentioned above, it is innate for humans to adjust the impedance and force of their arms skillfully while interacting with unknown environments [25]. From a biological point of view, humans can adapt their endpoint impedance of limbs through the central nervous system to achieve compliant interaction [26, 27]. Therefore, it would be crucial and significant to control robots in such a way for a variety of social applications such as health care, assembly lines and human-robot collaboration. In the literature, significant progress has been made to tackle such related issues [28–30]. Generally, it is of great importance to establish a desired impedance model for impedance control parameters which are difficult to choose from in practice. One of the most important parts of impedance control is to determine the desired impedance parameters. Nevertheless, most existing impedance control methods used fixed and manually designed impedance models. To improve the performance of compliant interaction under an unknown environment, optimization of the impedance model should be taken into consideration. In [31, 32], a desirable force regulation and tracking performance have been achieved using a linear quadratic regulator (LQR), which is operated by considering a minimal cost function, to optimize the impedance parameters under a vital precondition of completely known environment dynamics. However, such algorithms seem too conservative to achieve the desired robot-environment interaction since they lack the ability to incorporate unstructured and dynamic environments. Progress has been made in addressing such problems, and various optimal control schemes in the case of an unknown environment have been proposed in the literature. [33, 34] proposed a force sensorless admittance control for robots to tackle the unknown robot-environment interaction.

In this chapter, we propose a broad fuzzy neural control framework and apply it to a constrained manipulator which interacts with the unknown environment. Impedance adaptation is used to cope with the optimal problem of robot-environment interaction and BFNN is utilized to approximate the unknown dynamic model. Compared with the existing research, the major difficulties and contributions of this work are listed as follows.

1. A novel structure of BFNN, which can accommodate a new class of input vector by adjusting the membership and enhancement nodes adaptively, is proposed to improve the generalizing capability of FNN and tackle the difficulty of designing appropriate neural nodes for the NN controller. Compared with the recent work in [23] which also incorporates BLS in FNN, the major innovation of our work is that the membership and enhancements nodes are incremented adaptively rather than specifying beforehand so that the underlying issues can be resolved more effectively and reasonably.
2. Conventional NN or FNN adaptive controllers have shown a poor generalizing performance for different motions, e.g., once the orbit is changed, the controller should be returned to guarantee the desired tracking performance. In this work, an adaptive BFNN controller is designed to cope with such a problem by adapting the structure of BFNN in terms of the new input vector and historical information in neural nodes.
3. An impedance optimal adaptation scheme is developed based on [35] to improve the interaction performance between the robot and the environment. The novelty of such a scheme compared with [35] is that it tackles the problem of discontinuity caused by the mutation of control gain, by introducing an auxiliary soft function.
4. A tan-type Lyapunov function [36] is adopted to prevent the state overshoot during optimization of the impedance model and ensure the transient performance.

8.2 Problem Statement and Preliminaries

8.2.1 *Dynamic and Impedance Model*

In this chapter, robot-environment interaction and adaptive NN control are investigated by discussing the issues of state constraints, unknown plant model approximation and impedance control. Since the n-link manipulator is controlled in the Cartesian space, we consider the dynamic model based on *Lagrange-Euler* form as

$$M_x(q)\ddot{x} + C_x(q, \dot{q})\dot{x} + G_x(q) = \tau_x + \tau_e + \tau_f, \tag{8.1}$$

where $x \in \mathbb{R}^d$ is the end-effector's coordinates in Cartesian space; $q \in \mathbb{R}^k$ is the joint angle vector; d and k are the dimension of the Cartesian space and DOF of the manipulator; $\tau_x \in \mathbb{R}^d$ is the control torque regulated by the adaptive controller, $\tau_e \in \mathbb{R}^d$ denotes the impedance force at its end-effector while interacting with the environment and τ_f represents the external disturbance force; $M_x(q) \in \mathbb{R}^{d\times d}$ is a symmetric and positive definite inertia matrix of the manipulator; $C_x(q, \dot{q}) \in \mathbb{R}^{d\times d}$ and $G_x(q) \in \mathbb{R}^d$ represent the Coriolis and centrifugal forces vector and gravity torques, respectively, which are derived from joint space to Cartesian space as follows:

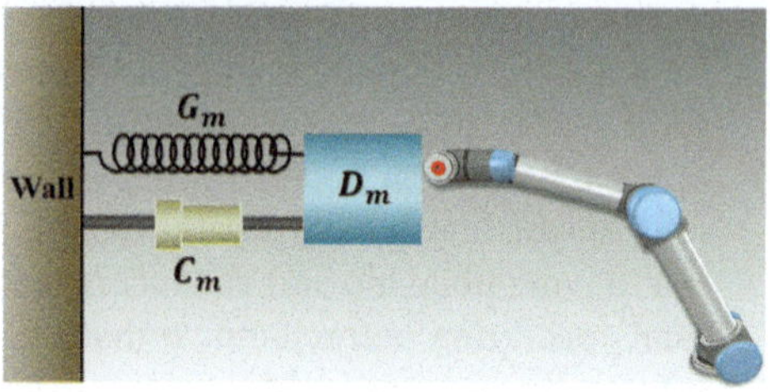

Fig. 8.1 Impedance model of robot-environment interaction

$$M_x(q) = J^{-T} M(q) J^{-1} \tag{8.2}$$
$$C_x(q, \dot{q}) = J^{-T} C(q, \dot{q}) J^{-1} - J^{-T} M(q) J^{-1} \dot{J} J^{-1} \tag{8.3}$$
$$G_x(q) = J^{-T} G(q), \tag{8.4}$$

where $J(q) \in \mathbb{R}^{k \times d}$ is a non-singular Jacobin matrix of the plant.

In this work, the manipulator is required to respond to a desirable compliance behaviour (Fig. 8.1) while contacting with the unknown environment in the presence of model uncertainties. The relationship between the target impedance model and the close-loop system can be defined as

$$M_m \ddot{x} + C_m \dot{x} + G_m (x - x_0) = -\tau_e, \tag{8.5}$$

where M_m, C_m and G_m denote the inertia, damping and stiffness matrices, $x_0 \in \mathbb{R}^d$ denotes the virtual desired reference trajectory which will equal to the desired reference trajectory x_d when there is no external force. When the manipulator works in free space, which means the position of the end-effector can follow the reference trajectory completely, it can be obtained that $x = x_0 = x_d, \forall t > 0$ such that $\tau_e = 0$.

8.2.2 *Adaptive Optimal Impedance Control*

In this chapter, we aim to design a computational optimal impedance controller, which can achieve favourable robot-environment interaction subjecting to the unknown environment model. The principle of optimal impedance adaptation proposed in [35] and its developed scheme are elaborated in this subsection.

Equation (8.5) has been proven for modelling the human limb under human-robot interaction [37]. However, since the mass matrix M_m is usually highly nonlinear and the research object of this work is robot-environment interaction, we will adopt a more simplified environment dynamic model to derive the optimal impedance target

$$C_m \dot{x} + G_m x = -\tau_e. \tag{8.6}$$

We first consider a linear state-space system

$$\dot{\zeta}(t) = A\zeta(t) + Bu(t), \tag{8.7}$$

where $\zeta = [x^T, s^T] \in \mathbb{R}^p$ denotes the state variable, $s \in \mathbb{R}^m$ is an intermediate state variable of the following system:

$$\begin{cases} \dot{s} = Ms \\ x_d = Ns \end{cases}, \tag{8.8}$$

where $M \in \mathbb{R}^{m\times m}$ and $N \in \mathbb{R}^{d\times m}$ are two tunable matrix. In this way, we can derive that $A = \text{diag}\{-G_m C_m^{-1}, M\}$, $B = [-C_m^{-1}, 0]^T$ and $u = \tau_e$ in terms of the environment dynamic model (8.6). The performance index function, which needs to be minimized by the system input $\tau_e = -K_m\zeta$, is defined as

$$\begin{aligned} J_o &= \int_0^\infty [(x - x_d)^T Q(x - x_d) + \tau_e^T R\tau_e]dt \\ &= \int_0^\infty [\zeta^T Q'\zeta + \tau_e^T R\tau_e]dt \end{aligned} \tag{8.9}$$

where $Q' = \begin{bmatrix} Q & -QN \\ -N^T Q & N^T QN \end{bmatrix}$. It is worth noting that the conventional optimal solution [38] by means of a well-known algebraic Ricotta equation (ARE) function is unavailable for obtaining the optimal control feedback gain, since the environment dynamic model is unknown. In this way, iterative learning is adopted to cope with this issue. To this end, some operators will first be defined as follows:

$$\begin{aligned} \bar{\zeta} &= [\zeta_1^2, \zeta_1\zeta_2, \ldots, \zeta_1\zeta_p, \zeta_2^2, \zeta_2\zeta_3, \ldots, \zeta_{p-1}\zeta_p, \zeta_p^2]^T \\ \Delta_{\bar{\zeta}} &= [\bar{\zeta}(t_1) - \bar{\zeta}(t_0), \bar{\zeta}(t_2) - \bar{\zeta}(t_1), \ldots, \bar{\zeta}(t_l) - \bar{\zeta}(t_{l-1})]^T \\ I_{\zeta\zeta} &= \left[\int_{t_0}^{t_1} \zeta \otimes \zeta dt, \int_{t_1}^{t_2} \zeta \otimes \zeta dt, \ldots, \int_{t_{l-1}}^{t_l} \zeta \otimes \zeta dt\right]^T \\ I_{\zeta\tau_e} &= \left[\int_{t_0}^{t_1} \zeta \otimes \tau_e dt, \int_{t_1}^{t_2} \zeta \otimes \tau_e dt, \ldots, \int_{t_{l-1}}^{t_l} \zeta \otimes \tau_e dt\right]^T, \end{aligned} \tag{8.10}$$

where ζ_i is the ith element of ζ, "$\otimes$" represents the Kronecker product.

Remark 8.1 In contrast to the impedance adaptation method in [35], which may lead to the discontinuity in control performance due to the mutation of control gain K_m, the novelty of our proposed scheme lies in that we introduce an auxiliary soft function which is defined as

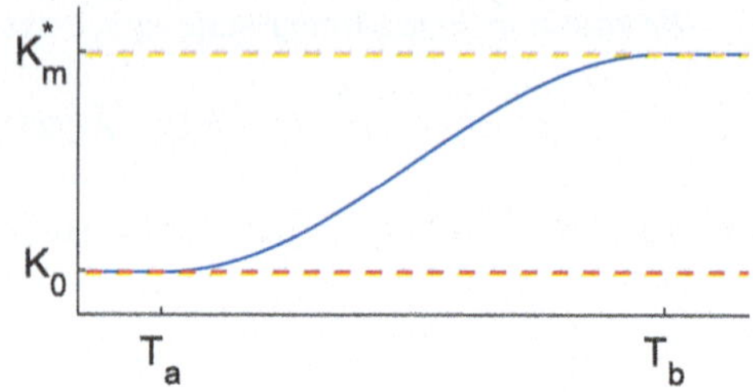

Fig. 8.2 Changing of the optimal impedance gain K'_m from K_0 to K^*_m

$$K'_m = \frac{K^*_m + K_0}{2} + \frac{K^*_m - K_0}{2} \sin\left(-\frac{\pi}{2} + \pi \frac{t}{T_s}\right), \tag{8.11}$$

where K_0 and K^*_m are the initial and optimal impedance gain, respectively. As shown in Fig. 8.2, $T_s = T_b - T_a$ denotes the sampling period. According to the property of the sin function, the control gain K_m is changing smoothly and monotonically from K_0 to K^*_m which can prevent the discontinuity of the controller.

To proceed, the principle of the improved impedance optimal scheme is summarized in Algorithm 8.1.

Algorithm 8.1 Framework of adaptive impedance optimal scheme

Input: The set of initial feedback gain K_0 with exploration noise υ_m; The set of state variable ζ;
Output: Optimal feedback gain K^*_m;
1: phase 1: Set $\tau_{e0} = K_0\zeta + \upsilon_m$ as the initial input torque while the manipulator is contacting with the environment;
2: **repeat**
3: compute operates $\Delta_{\bar{\zeta}}$, $I_{\zeta\zeta}$ and $I_{\zeta\tau_e}$;
4: **until** (rank$[I_{\zeta\zeta}, I_{\zeta\tau_e}] = \frac{p(p+1)}{2} + pd$)
5: **repeat**
6: phase 2: Set $\tau_{e0} = K_0\zeta$
7: Solve P_m and K_{m+1} according to

$$\begin{bmatrix} \hat{P}_k \\ \text{vec}(K_{m+1}) \end{bmatrix} = (\Xi^T_m \Xi_m)^{-1} \Xi^T_m F_m \tag{8.12}$$

8: let $m + 1 \rightarrow m$;
9: **until** $\|P_m - P_{m-1}\| \leqslant \varepsilon_P$
10: Sample the impedance gain K'_m from K_0 to K^*_m during the sampling period T_s
11: **repeat**
12: phase 3: Set $\tau_{e0} = K'_m\zeta$
13: **until** $t > T_b$
14: **return** K^*_m;

Specifically, the following definition should be supplemented in step 7:

$$\begin{aligned}\hat{P}_k &= [P_{11}, 2P_{12}, P_{22}, 2P_{23}, \ldots, 2P_{p(p-1)}, P_{pp}] \\ \Xi_m &= [\Delta_{\bar{\zeta}}, -2I_{\zeta\zeta}(I_p \otimes K_m^T R) - 2I_{\zeta\tau_e}(I_p \otimes R)] \\ F_m &= -I_{\zeta\zeta}\text{vec}(Q_m),\end{aligned} \tag{8.13}$$

where I_p is a p-dimension unit matrix, P_{ij} denotes the element of P_m, "vec(·)" is an operator which aims to stretch the matrix from $\mathbb{R}^{n\times m}$ to $\mathbb{R}^{1\times nm}$, Q_m can be obtained from Q where $Q_m = Q + K_M^T R K_M^T$.

8.2.3 Broad Learning System

Recently, a novel incremental learning called a broad learning system has been proposed in literature [20]. Notably, it is designed by the inspiration of random vector functional-link neural networks and has been well demonstrated its generation and efficient approximation ability in recognition and classification fields. The major difference between BLS and the conventional RVFLNN is that the input vector of the hidden layer has been replaced by a set of feature mappings. Particularly, a prominent generating capability can be achieved by producing the enhancement nodes in terms of the feature mappings.

The schematic diagram of the broad learning system is shown clearly in Fig. 8.3. In this framework, X and $Y \in \mathbb{R}^c$ denote the input and output matrix, respectively. Next, the n feature mappings F_n will be derived from the input matrix X

$$F_i = \psi(XW_{f_i} + \beta_{f_i}), \qquad i = 1\ldots n \tag{8.14}$$

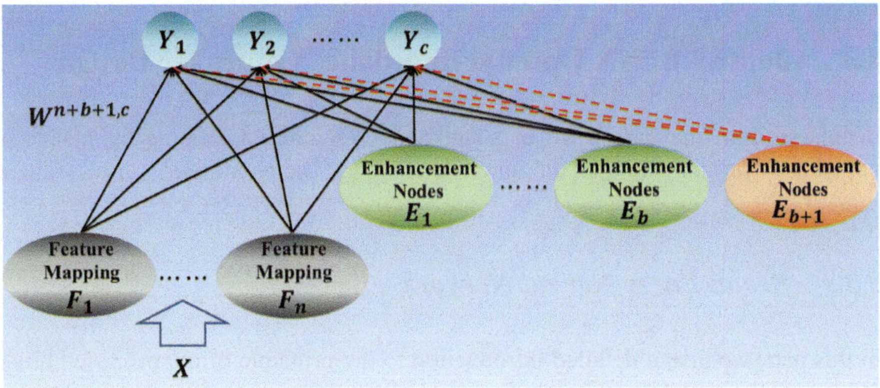

Fig. 8.3 Schematic diagram of BLS

where the neural weight W_{f_i} and biases β_{f_i} are chosen randomly and then kept constant during training. Then, the enhancement nodes are represented in terms of the feature mappings as

$$E_j = \xi(F^n W_{e_j} + \beta_{e_j}), \qquad j = 1 \ldots b \tag{8.15}$$

Here, both $\psi(\cdot)$ and $\xi(\cdot)$ are the transfer functions. Without loss of generality, the output of a broad learning neural network is calculated as

$$\begin{aligned} Y &= [F_1, \ldots, F_n | \xi(F^n W_{e_1} + \beta_{e_1}), \\ &\quad \ldots, \xi(F^n W_{e_b} + \beta_{e_b})] W^{n+b,c} \\ &= [F_1, \ldots, F_n | E_1, \ldots, E_b] W^{n+b,c} \\ &= [F^n | E^b] W^{n+b,c} \\ &= G^{n+b,c} W^{n+b,c}. \end{aligned} \tag{8.16}$$

From the supervised learning point of view, the recognition problem can be well tackled by BLS since the neural network can be trained efficiently by solving the pseudo-inverse of $[F^n|E^b]$, i.e., $W^{n+b,c} = [F^n|E^b]^+ Y$, whose output vectors are usually given beforehand. Aiming at improving the approximation accuracy of the neural network, it can produce extra enhancement nodes into the original network to extract more feature information from the input data. In this way, the augmented layer is denoted as

$$\begin{aligned} G^{b+1} &= [G^{n+b,c} | E_{b+1}] \\ &= [G^{n+b,c} | \xi(Z^n W_{e_{j+1}} + \beta_{e_{j+1}})], \end{aligned} \tag{8.17}$$

where $W_{e_{j+1}}$ and $\beta_{e_{j+1}}$ are also generated randomly.

8.3 Adaptive BFNN Optimal Impedance Controller Design

In this section, the design of an adaptive BFNN control scheme by means of impedance learning under an unknown environment is elaborated.

8.3.1 Broad Fuzzy Neural Network

In this part, we give a detailed introduction to the principle of the proposed broad fuzzy neural network framework. In a conventional fuzzy neural network, the neural nodes and fuzzy rules are usually designed manually in terms of the expertise experience which seems to be highly subjective. When the desired task can not be

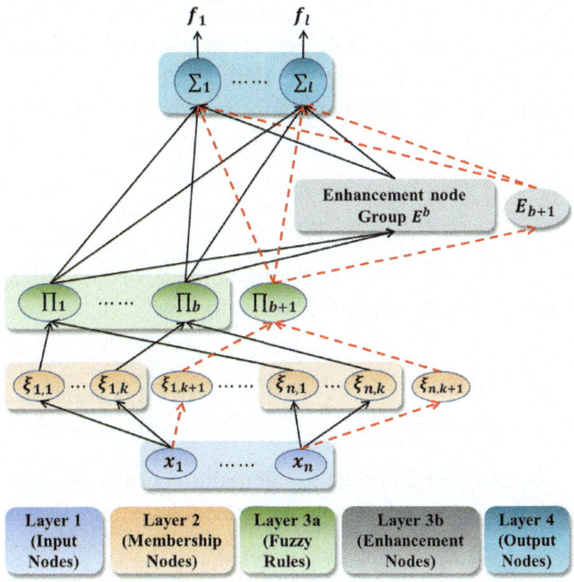

Fig. 8.4 Schematic diagram of broad fuzzy neural network

learned well through the designed network model, general actions are either tuning the network parameters or increasing the neural nodes of the network. It can be seen that a desired network model can only be obtained by trial and error repeatedly and inefficiently. For these reasons, this part will bring forward a solution determining the number and position of neural nodes by incorporating BLS.

Figure 8.4 illustrates the principle of proposed BFNN which has been shown the excellent approximation characteristic in the last subsection. Obviously, there are four layers in this framework. Layer 1 contains the network input vectors x_n which are used to derive the membership nodes in layer 2. From the robotic neural control point of view, we choose the Gaussian exponential function as the transfer membership function

$$\xi_{i,j}(x_i) = \exp\left[\frac{-(x_i - c_{i,j})^T(x_i - c_{i,j})}{\eta_w^2}\right], \tag{8.18}$$

where $\xi_i = [\xi_{i,1}, \xi_{i,2}, \ldots, \xi_{i,k}]^T$, $c_{i,j}$ and η_w denote the centre and width of the membership function, respectively.

Notably, we choose the initial number of membership nodes as $k = 1$. During the network training, the membership nodes can be incremented adaptively according to the second norm distance between Gaussian centres and input variables. To be specific, the norm distance is defined as

$$d(x, \overline{c}_{min}) = \|x - \overline{c}_{min}\|, \tag{8.19}$$

where $\overline{c}_{min} = (\sum_{i=1}^{m} c_{min_i})/m$ denotes the average position from the m closet centers to the input vector x. Once the norm distance d exceeds the predefined threshold value Θ, a new membership node will be produced. The parameters of incremental nodes and the new weight of enhancement nodes are defined as

$$c_{i,b+1} = \overline{c}_{min} + \beta_1(x - \overline{c}_{min}) \tag{8.20}$$
$$W^{b+1} = [W_b|0, 0], \tag{8.21}$$

where $\beta_1 \in \mathbb{R}^k$ is the tunable parameter. Next, the fuzzy rules in layer 3a are calculated by n-norm product operation as follow:

$$\varsigma_j = \prod_{i=1}^{n} \xi_{i,j}(x_i), \tag{8.22}$$

Aiming at reducing the calculation burden, a trigonometric expansion scheme [39] will be employed to generate the enhancement nodes. In this way, the enhancement vector is defined as

$$\begin{aligned} E^b &= [E_1, E_2, \ldots, E_b] \\ &= [cos(\varsigma_1), sin(\varsigma_1), \ldots, cos(\varsigma_b), sin(\varsigma_b)]. \end{aligned} \tag{8.23}$$

Hence, the continuous function approximated by BFNN can be represented as

$$\begin{aligned} f(x) &= [W_b|W_{b+1}]^T[\varsigma_b, \varsigma_{b+1}|E^b, E_{b+1}] \\ &= W^T\vartheta. \end{aligned} \tag{8.24}$$

Remark 8.2 Aiming at improving the generalizing capability, the superiority of the BFNN framework is able to accommodate any new sample from input without retuning the network's parameters. In contrast to the recent research [23], the number of membership and enhancements nodes is augmented in terms of the input vector and history information of the network adaptively instead of specifying beforehand. Additionally, it is worth noting that such a framework tends to be more suitable for a control point of view than BLS since the orthogonal calculation in BLS takes place by the Gaussian exponential function.

8.3.2 *Adaptive BFNN Controller Design*

In this part, we design the adaptive BFNN controller with a detailed presentation of the design process and formative stability analysis. Figure 8.5 shows the configuration of the proposed controller. In general, the Cartesian tracking error signals are first defined as

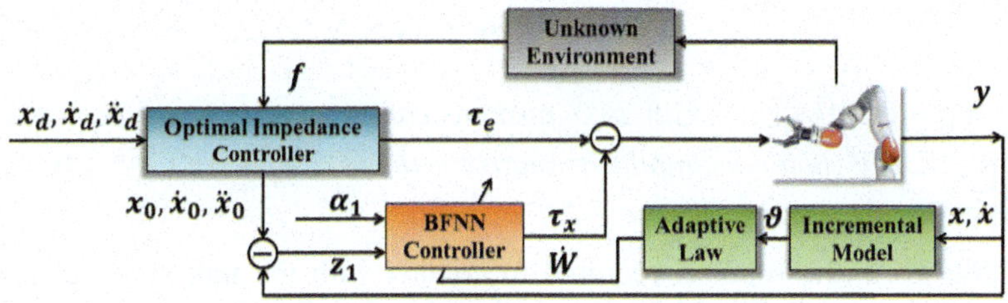

Fig. 8.5 Control diagram of the BFNN controller

$$z_1 = x - x_0 \tag{8.25}$$
$$z_2 = \dot{x} - \alpha_1, \tag{8.26}$$

where α_1 is the virtual controller which will be discussed hereinafter. For the convenience of analysis, the dynamic model (8.1) is transferred to the state-space function

$$\begin{aligned} \dot{x}_1 &= x_2 \\ \dot{x}_2 &= M_x^{-1}[\tau_x - \tau_e - C_x(q, \dot{q})x_2 - G_x(q)], \end{aligned} \tag{8.27}$$

where $x_1 = x$ and $x_2 = \dot{x}$. Practically, an excessive overshoot of position and velocity, which may occur during the interaction with the unknown environment, can lead to unpredictable and undesirable results for the robotic manipulator. For these reasons, we introduce a Barrier Lyapunov Function [36] to tackle the problem related to the state constraints:

$$V = \frac{\iota^2}{\pi} \tan\left(\frac{\pi \kappa^2}{2\iota^2}\right), \tag{8.28}$$

where κ is the system state and it will be restricted with $\|\kappa\| < \iota$. Also, it can be derived that (8.28) will transfer to the following quadratic form in terms of L'Hospital's rule once the constraint is omitted:

$$\lim_{\iota \to \infty} \frac{\iota^2}{\pi} \tan\left(\frac{\pi \kappa^2}{2\iota^2}\right) = \frac{1}{2}\kappa^2, \tag{8.29}$$

Lemma 8.1 [40] *Consider a continuous Lyapunov function $V(t) > 0$ with $V(0)$ is bounded and $\forall t \in \mathbb{R}^+$. It can be concluded that $V(t)$ is bounded if the derivative of the Lyapunov function satisfies the following inequality:*

$$\dot{V}(t) \leqslant -\varpi_1 V(t) + \varpi_2,$$

where ϖ_1 and ϖ_2 are tunable positive constants. In the following, we specify the virtual control α_1 in (8.25) as the following form to enable the Cartesian tracking error z_1 remain constraint, i.e.,$|z_{1i}| < \iota_i$, $(i = 1, 2, \ldots, n)$:

$$\alpha_1 = -K_1\Phi + \dot{x}_0, \tag{8.30}$$

where $K_1 = \text{diag}\{k_{11}, \ldots, k_{1n}\}$ *is a positive constant matrix with* $k_{ln} > 0$, $\Phi = [\phi_1, \phi_2, \ldots, \phi_n]^T$ *is a nonsingularity matrix with* $\phi_n = \frac{\iota_n^2}{2\pi z_{1n}} sin\left(\frac{\pi z_{1n}^2}{\iota_n^2}\right)$ *in terms of* L'Hospital's *rule.*

In this work, the robot is required to contact with the unknown environment through the proposed algorithm. Hence, the adaptive control torque is designed by taking the contact force τ_e into account which can be derived in subsection II-B as the following form:

$$\begin{aligned}\tau_x = &-\begin{bmatrix} \frac{z_{11}}{\cos^2\left(\frac{\pi z_{11}^2}{2\iota_1^2}\right)} \\ \vdots \\ \frac{z_{1n}}{\cos^2\left(\frac{\pi z_{1n}^2}{2\iota_n^2}\right)} \end{bmatrix} + M_x + C_x + G_x - K_p z_2 \\ &- K_r\,\text{sgn}(z_2) + \tau_e,\end{aligned} \tag{8.31}$$

where $\hat{M}$, $\hat{C}$ and $\hat{G}$ are the estimates of robotic dynamic models which are approximated by BFNN (8.24) as

$$D = W_D^{*T}\vartheta_D + \epsilon_D \tag{8.32}$$
$$C = W_C^{*T}\vartheta_C + \epsilon_C \tag{8.33}$$
$$G = W_G^{*T}\vartheta_G + \epsilon_G, \tag{8.34}$$

where ϵ_D, ϵ_C and ϵ_G are the approximation errors.

Here, it is worth to know that $\tilde{(\cdot)} = \hat{(\cdot)} - (\cdot)^*$. Substituting the control torque (8.31) into the robotic model (8.1), the error dynamic is calculated as

$$\begin{aligned}M\dot{z}_2 = &-\begin{bmatrix} \frac{z_{11}}{\cos^2\left(\frac{\pi z_{11}^2}{2\iota_1^2}\right)} \\ \vdots \\ \frac{z_{1n}}{\cos^2\left(\frac{\pi z_{1n}^2}{2\iota_n^2}\right)} \end{bmatrix} - Cz_2 + \tilde{D}_x\dot{\alpha}_1 + \tilde{C}_x\alpha_1 + \tilde{G}_x \\ &- K_p z_2 - K_r\,\text{sgn}(z_2) + E,\end{aligned} \tag{8.35}$$

where $E = \epsilon_D\dot{\alpha}_1 + \epsilon_C\alpha_1 + \epsilon_G$ with $K_r > \|E\|$.

Remark 8.3 For the parameters of the adaptive BFNN controller, the state constraint is positive $\iota_i > 0$. The control parameters $K_1 = \text{diag}\{k_{11}, \ldots, k_{1n}\}$, $K_p = \text{diag}\{k_{p1}, \ldots, k_{pn}\}$ and $K_r = \text{diag}\{k_{r1}, \ldots, k_{rn}\}$ are designed through trial and error with $k_{1n} > 0$, $k_{pn} > 0$ and $k_{rn} > 0$ so as to obtain a stable and desired close-loop control system. The norm distance d of BFNN determines the number of membership and enhancements nodes. Small d results in more nodes and vice versa.

8.3.3 Stability Analysis

In this part, the Lyapunov approach is used to rigorously evaluate the stability of the control system and a stable adaptive updated law of BFNN (8.24) will be derived in detail.

Theorem 8.1 *Taking the manipulator dynamic system (8.27) with control torque (8.31), state constraints (8.28) and the broad learning scheme (8.20)–(8.24) into consideration, the proposed incremental learning control scheme can guarantee that the tracking errors are bounded and can converge to the compact set* $\Omega_{z_1} := z_1 \leqslant \sqrt{2(V_1(0) + \frac{\nu}{\gamma})}$ *with suitable control parameters. Furthermore, all the signals in the system remain uniformly ultimately bounded (UUB) during the incremental training.*

Proof Considering the following Lyapunov function:

$$\begin{aligned} V_1(t) = &\sum_{i=0}^{n} \frac{\iota_i^2}{\pi} \tan\left(\frac{\pi z_{1i}^2}{2\iota_i^2}\right) + \frac{1}{2} z_2^T D z_2 \\ &+ \frac{1}{2}\tilde{W}_D^T \tilde{W}_D + \frac{1}{2}\tilde{W}_C^T \tilde{W}_C + \frac{1}{2}\tilde{W}_G^T \tilde{W}_G. \end{aligned} \tag{8.36}$$

Upon taking the derivative of the candidate function using the error dynamic (8.35) and using the skew-symmetric matrix $[\dot{D} - 2C]$, we have

$$\begin{aligned} \dot{V}_1(t) = &-\sum_{i=0}^{n} \frac{k_{1i}\iota_i^2}{\pi} \tan\left(\frac{\pi z_{1i}^2}{2\iota_i^2}\right) - z_2^T K_p z_2 \\ &+ z_2^T (E - K_r \operatorname{sgn}(z_2)) \\ &+ z_2^T (\tilde{W}_D^T \vartheta_D \dot{\alpha}_1 + \tilde{W}_C^T \vartheta_C \alpha_1 + \tilde{W}_G^T \vartheta_G) \\ &+ \tilde{W}_D^T \dot{\hat{W}}_D + \tilde{W}_C^T \dot{\hat{W}}_C + \tilde{W}_G^T \dot{\hat{W}}_G. \end{aligned} \tag{8.37}$$

The adaptive neural weight updated laws are chosen as

$$\dot{\hat{W}}_D = -\vartheta_D \dot{\alpha}_1 z_2 - \beta_D \hat{W}_D \tag{8.38}$$

$$\dot{\hat{W}}_C = -\vartheta_C \alpha_1 z_2 - \beta_C \hat{W}_C \tag{8.39}$$

$$\dot{\hat{W}}_G = -\vartheta_G z_2 - \beta_G \hat{W}_G, \tag{8.40}$$

where $\beta_D, \beta_C, \beta_G$ are the robust items of the BFNN. The differential Lyapunov function (8.37) can be further expressed as follows by substituting the update laws (8.38)–(8.40) into (8.37):

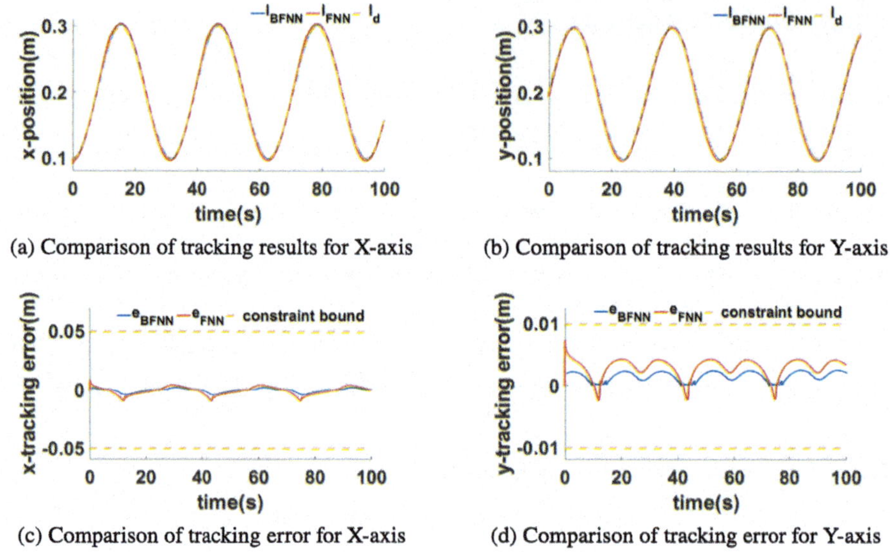

(a) Comparison of tracking results for X-axis
(b) Comparison of tracking results for Y-axis
(c) Comparison of tracking error for X-axis
(d) Comparison of tracking error for Y-axis

Fig. 8.6 Comparison of tracking results. l_{BFNN} and e_{BFNN} are the tracking trajectory and error respectively with the BFNN controller. l_{FNN} and e_{FNN} are the tracking trajectory and error respectively with the RBFNN controller. l_d is the reference trajectory

$$\begin{aligned}\dot{V}_1(t) = &-\sum_{i=0}^{n}\frac{k_{1i}\iota_i^2}{\pi}\tan\left(\frac{\pi z_{1i}^2}{2\iota_i^2}\right) - z_2^T K_p z_2 \\ &-\beta_D \tilde{W}_D^T \hat{W}_D - \beta_C \tilde{W}_C^T \hat{W}_C - \beta_G \tilde{W}_G^T \hat{W}_G.\end{aligned} \tag{8.41}$$

Taking the Young's inequality $-\tilde{W}^T W^* \leqslant \frac{1}{2}(\tilde{W}^T\tilde{W} + W^{*T}W^*)$ and the equality $-\tilde{W}^T\hat{W} = -\tilde{W}^T(W^* + \tilde{W})$ into account, yields

$$\begin{aligned}\dot{V}_1(t) \leqslant &-\sum_{i=0}^{n}\frac{k_{1i}\iota_i^2}{\pi}\tan\left(\frac{\pi z_i^2}{2\iota_i^2}\right) - z_2^T K_p z_2 \\ &-\frac{1}{2}(\tilde{W}_D^T\tilde{W}_D + \tilde{W}_C^T\tilde{W}_C + \tilde{W}_G^T\tilde{W}_G) \\ &+\frac{1}{2}(W_D^{*T}W_D^* + W_C^{*T}W_C^* + W_G^{*T}W_G^*) \\ \leqslant &-\gamma V_1(t) + \nu,\end{aligned} \tag{8.42}$$

where $\nu = \frac{1}{2}(W_D^{*T}W_D^* + W_C^{*T}W_C^* + W_G^{*T}W_G^*)$, $\gamma = \min(2k_{11}, 2k_{12}, \ldots, 2k_{1n}, \frac{2\lambda_{\min}(K_p)}{\lambda_{\max}(D)})$ with λ respects the eigenvalue of $(\cdot)$. Aiming at guaranteeing the differential Lyapunov candidate being negative definite, the control parameters should be specified as $|K_p| > 0$ and $|K_1| > 0$ in this work. For later development, we multiply (8.42) by $e^{\gamma t}$ such that

$$\frac{d(V_1(t)e^{\gamma t})}{dt} \leqslant \nu e^{\gamma t}. \tag{8.43}$$

To solve the integral of the above inequality (8.43), we have

$$\begin{aligned} V_1(t) &\leqslant e^{-\gamma t} V_1(0) + \frac{\nu}{\gamma}(1 - e^{-\gamma t}) \\ &\leqslant V_1(0) + \frac{\nu}{\gamma}. \end{aligned} \tag{8.44}$$

According to the above discussion in terms of z_1, we can get

$$\frac{1}{2} z_1^T z_1 \leqslant V_1(0) + \frac{\nu}{\gamma}. \tag{8.45}$$

Particularly, we can derive that $\|z_1\| \leqslant \sqrt{2(V_1(0) + \frac{\nu}{\gamma})}$. Thus, the control signal (8.31) even the dynamics system (8.27) are bounded by considering (8.43)-(8.45). The tracking error z_1 converges to the specific compact set Ω_{z_1} in terms of (8.45). It can be further demonstrated that the tracking errors satisfy the requirement of uniformly ultimately bounded. Simultaneously, choosing suitable design parameters can guarantee the desired performance of compliant interaction. In this way, we can conclude that the control system is in the sense of Lyapunov stable by adopting the proposed algorithm.

8.4 Simulation

To verify the incremental learning and interaction capability of the proposed algorithm, two comparative simulations are carried out by utilizing a two-DOF manipulator with a force sensor installed at the end-effector. In this section, simulations are performed with the help of the Robotics Toolbox of Matlab [41] and the manipulator is required to interact with the unknown environment which contains both free and constraint movement. The parameters of the manipulator's dynamic model are shown in Table 8.1.

8.4.1 Case 1. Evaluation of Tracking Performance

In this case, the manipulator first tracks the desired circle which is defined as

$$x_d(t) = \begin{bmatrix} 0.2 - 0.1\cos(0.2t) \\ 0.2 + 0.1\sin(0.2t) \end{bmatrix}, \tag{8.46}$$

Table 8.1 Parameters of the manipulator

Parameter	Description	Value	Unit
m_1	Mass of link 1	2.0	kg
m_2	Mass of link 2	2.0	kg
l_1	Length of link 1	0.2	m
l_2	Length of link 2	0.2	m
I_1	Inertia of link 1	0.027	kg m^2
I_2	Inertia of link 2	0.027	kg m^2
g	Gravity acceleration	9.8	m/s^2

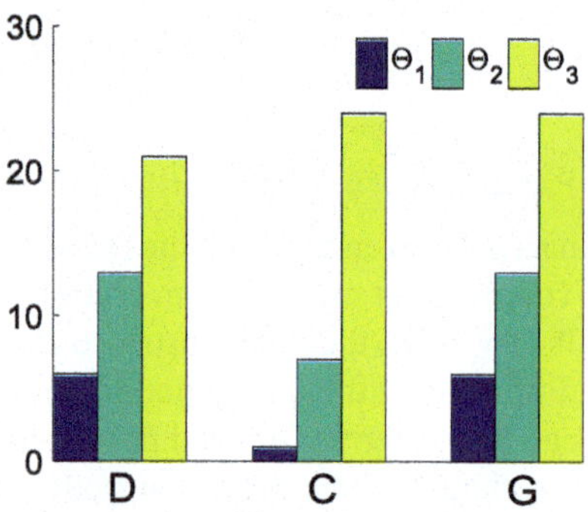

Fig. 8.7 The number of incremental nodes in BFNN with different Θ: D is the number of c_D. C is the number of c_C. G is the number of c_G

whose centre is located at [0.2, 0.2] m with a 0.1 m radius and it will move from the initial Cartesian position of [0.125, 0.2]m. During the tracking process, the adaptive BFNN constraint controller which is designed in (8.31) is utilized to guarantee transient and steady performance. For the proposed controller, the constraint parameter is chosen as $\iota_n = [0.05, 0.01]$ and the control gains are specified as $K_1 = \text{diag}\{80, 80\}$, $K_r = \text{diag}\{5, 5\}$ and $K_p = \text{diag}\{60, 60\}$. The parameters of the BFNN updated law are chosen as $\beta_D = \beta_C = \beta_G = 5$. The distance threshold of incremental regulation is designed as $\Theta = [\Theta_D, \Theta_C, \Theta_G] = [0.1, 0.3, 0.1]$ which denotes the threshold for neural node c_D, c_C and c_G, respectively.

Furthermore, a conventional FNN [16], whose control parameters are designed identically as the above controller, is employed to make a comparison for the proposed network structure. It also adopts the exponential function as the membership function and its neural centres of membership function in this case are chosen as $c'_D = [0.1, 0.2, 0.3] \times [0.1, 0.2, 0.3]$, $c'_C = [0.1, 0.2, 0.3] \times [0.1, 0.2, 0.3] \times [0.01, 0.02, 0.03] \times [0.01, 0.02, 0.03]$ and $c'_G = [0.1, 0.2, 0.3] \times [0.1, 0.2, 0.3]$ with a total of $3^2 + 3^4 + 3^2 = 99$ neural nodes. The simulation results are shown in Fig. 8.6. As shown in Fig. 8.8c and d, both structures can satisfy the state constraints condition. However, the BFNN controller may lead to smaller tracking errors, which also verifies the theoretical analysis in Theorem 1. Thus, we can easily conclude that the proposed controller outperforms the RBFNN controller.

In the process of control learning, the design of distance threshold Θ plays an important role in the BFNN approximator. The distance threshold determines whether the new neural centre c is supposed to increment or not. Figure 8.7 shows the simulation results about the number of incremental neural nodes under three different groups of Θ ($\Theta_1 = [0.1, 0.3, 0.1]$, $\Theta_2 = [0.05, 0.1, 0.05]$ and $\Theta_3 = [0.03, 0.05, 0.03]$). The number of neural nodes increases with the decrease of Θ. Accordingly, BFNN can result in more desirable tracking control performance than the conventional FNN controller using less number of neural nodes.

8.4.2 Case 2. Evaluation of Robustness Performance

In addition to motion control accuracy, another key characteristic in manipulation control performance is robustness. In practical terms, a system without sufficient robustness capability may lead to performance degradation even fatal results in the presence of external disturbance. Previous simulation results have demonstrated the superior motion-tracking performance of the BFNN controller. In this part, we focus on the comparison of robustness in the presence of external disturbance among BFNN, FNN controller designed above and H_∞ controller [42] which is competent to improve the robustness of the robot model on motion tracking. To proceed, a friction model relative to joint velocity is designed to represent the external disturbance force in (8.1) as

$$\tau_f = -J^{-T}(a * sign(\dot{q}) + b * \dot{q}), \tag{8.47}$$

where $a = \text{diag}\{2, 2\}$ and $b = \text{diag}\{0.2, 0.2\}$. The tracking performance and tracking error of the comparison simulations are displayed in Fig. 8.8. Notably, the proposed BFNN controller has less tracking error under the motion friction which can be concluded that the proposed method has superior robust capability to eliminate the effect of external disturbance.

8.4.3 Case 3. Evaluation of Interaction Capability

Robot-environment interaction is regarded as one of the most vital capabilities of a robot. In this part, a convincing comparative simulation is set up to illustrate the effectiveness between the improved impedance optimal scheme and the conventional LQR optimal method. Both of these simulations employ the adaptive BFNN controller in case 1 equally. The simulation scenario is shown in Fig. 8.9.

The manipulator is first required to track the circle (8.46) and then it will contact the unknown object whose x-axis is located at $x_{ob} = 0.22$m. For ease of analysing the capability of interaction but without loss of generality, we assume that the interacting force exerted by the object is only in the X direction, however, the surface of the object

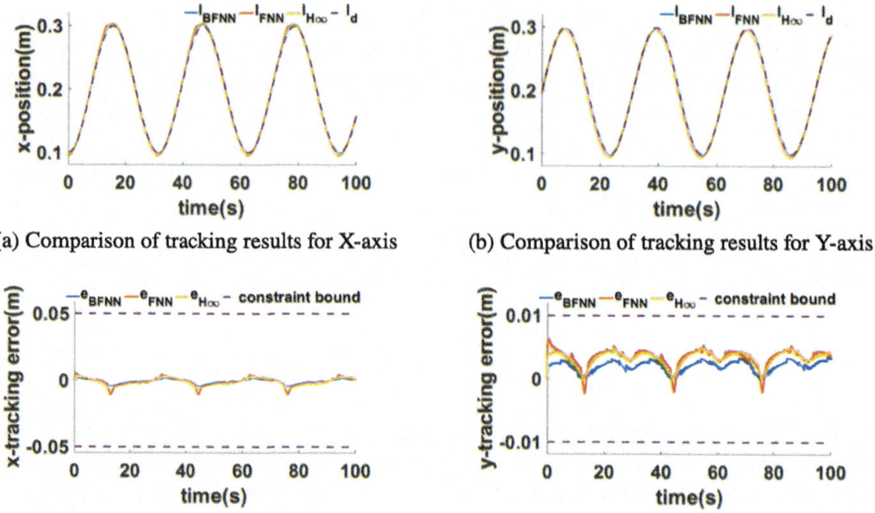

(a) Comparison of tracking results for X-axis (b) Comparison of tracking results for Y-axis

(c) Comparison of tracking error for X-axis (d) Comparison of tracking error for Y-axis

Fig. 8.8 Comparison of tracking results under external disturbance. l_{BFNN} and e_{BFNN} are the tracking trajectory and error respectively with the BFNN controller. l_{FNN} and e_{FNN} are the tracking trajectory and error respectively with the RBFNN controller. $l_{H\infty}$ and $e_{H\infty}$ are the tracking trajectory and error respectively with H_∞ controller. l_d is the reference trajectory

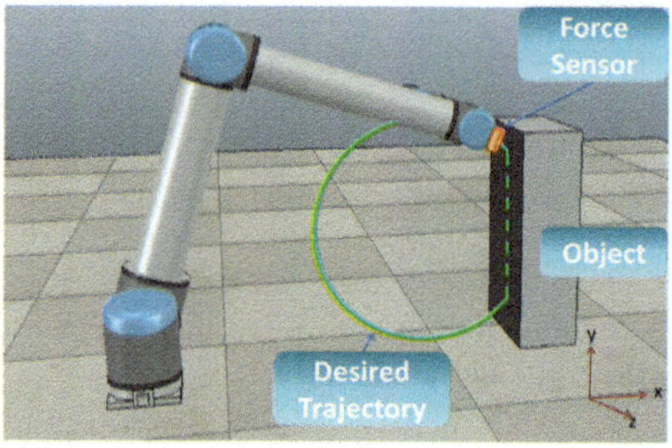

Fig. 8.9 Simulation scenario: end-effector of manipulator interacts with an unknown object

is smooth perfectly such that there is no external friction along the Y direction. Next, we establish the object's impedance model as $-f = [0.01\dot{x} + 10(x - 0.22), 0]$. The parameters of the cost function (8.9) are set as $N = 0.05$, $Q = 70$ and $R = 5$. The initial impedance gain is chosed as $K_0 = [-30.01, 1.47]$ and the well known exploration noise is design as $\upsilon_m = \sum_{k=1}^{8} \frac{0.05}{k} sin(kt)$. Then, the impedance optimal scheme is conducted according to Algorithm 8.1 and it will stop while $\|P_m - P_{m-1}\| < \varepsilon_P = 10^{-3}$. To make a meaningful comparison, the conventional LQR optimal method is adopted in this simulation whose state-space model is specified as $G_m = 1$, $C_m = 0.01$ and $M = -1$ in (8.7).

Fig. 8.10 Comparison of interaction tracking results. a l_4 is the reference trajectory. l_1, l_2 and l_3 are the tracking trajectories by using improved impedance optimal scheme, LQR and optimal method in [35] respectively. b e_1 and e_2 are the tracking errors corresponding to the virtual desired trajectory x_0 by using an improved impedance optimal scheme and LQR respectively

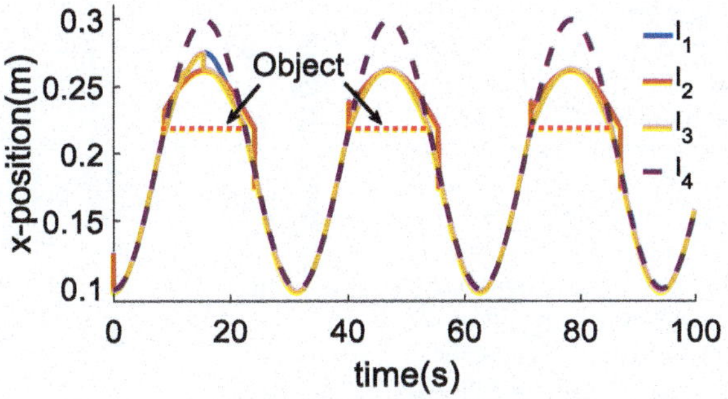

(a) Interaction tracking results for X-axis

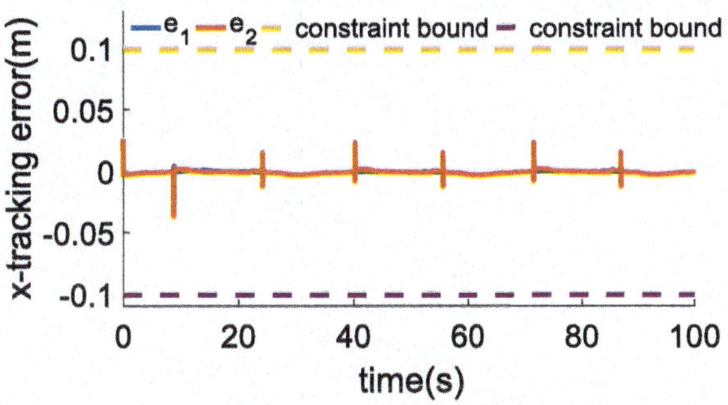

(b) Interaction tracking error for X-axis

As a result, the simulation of tracking performance is shown in Fig. 8.10. In this case, less oscillatory is achieved and discontinuity is omitted apparently due to the soft function in contrast to LQR and the optimal method in [35]. In addition, the tracking state constraints can also be satisfied during the interaction. Figure 8.11 shows the evolution process of the impedance optimization. Synthesizing the simulation of Figs. 8.10 and 8.11, it is obvious that the proposed scheme can also achieve the favourable interaction under the same initial parameters and state condition without the environment model knowledge such that it can be applied in more fields than LQR.

8.5 Experiment

In this section, the performance of the optimal interaction between the robot and environment is demonstrated by experimenting with a 7-DOF Baxter robot manipulator as shown in Fig. 8.12. The manipulator of Baxter contains several joint sensors whose resolution is 14 bits in the range of 360 degrees. During the task operation, the first four joints can offer 0–50 Nm torques and the last three joints can provide 0–15

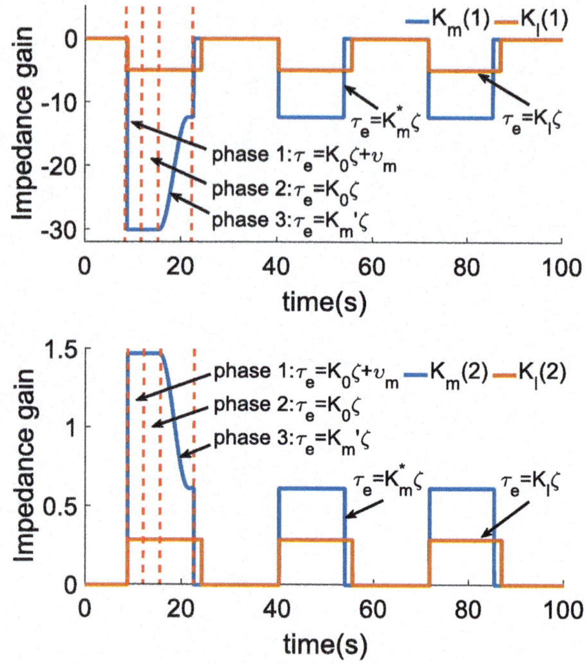

Fig. 8.11 Evolution of the impedance gain: K_l and K_m are optimized by LQR and proposed method respectively. Phases 1, 2, and 3 in Fig. 8.10 correspond to the 1, 6, 12 steps in the Algorithm 8.1, respectively. Phase 1 is the initial stage of the whole scheme. Phase 2 is the process of seeking the optimal impedance gain K_m^*. Phase 3 is the transfer stage from adopting initial impedance gain K_0 to K_m^* smoothly

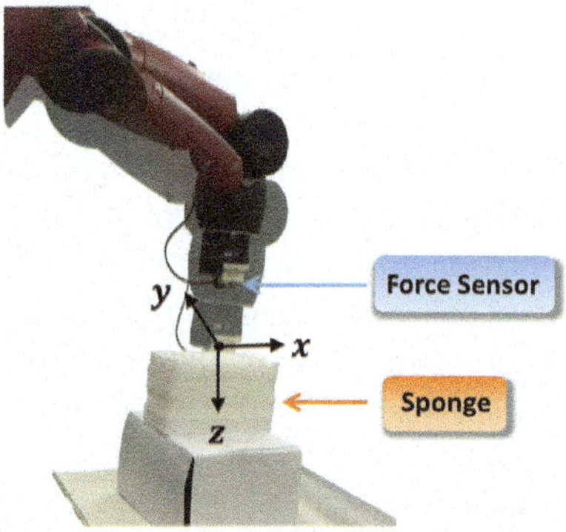

Fig. 8.12 Experimental scenario: Baxter robot interacts with sponge

Nm torques. The manipulator, which is equipped with the Mini45 Force/Transducer Sensor, is required to contact with sponge along with the z-axis to verify the flexibility of the end-effector by using the proposed algorithm. The force sensor and the system controller are communicated by UDP protocol whose sampling rate and control rate are set as 100 Hz and 50 Hz, respectively.

In this experiment, we assume that the initial position of the end-effector is located at $[0.77, -0.27, 0.35]$m and then it will contact with the sponge along the desired trajectory as

$$x_d(t) = \begin{bmatrix} 0.77 \\ -0.27 \\ 0.3\exp(-t) + 0.05 \end{bmatrix}. \tag{8.48}$$

While tracking the desired trajectory (8.48), the proposed adaptive BFNN controller is executed to guarantee the transient control performance by choosing the parameters as $\iota_n = [0.1, 0.1, 0.1]$, $K_1 = \text{diag}\{23, 23, 35\}$, $K_r = \text{diag}\{3, 3, 3\}$, $K_p = \text{diag}\{50, 50, 75\}$, $\beta_D = \beta_C = \beta_G = 5$, $\Theta = [\Theta_D, \Theta_C, \Theta_G] = [0.1, 0.5, 0.1]$. To obtain the optimal impedance model of the environment, an experimental comparison between the improved optimal scheme and the traditional LQR method is considered. In this part, the weights of the cost function in (8.9) are specified as $N = 1$, $Q = 370$ and $R = 0.003$ for both two cases. For our improved optimal algorithm, the iteration of optimization will stop while $||P_m - P_{m-1}|| < 0.1$. Moreover, the LQR method requires that the model of the environment is known such that the desired impedance model can be derived from the Riccati equations. Particularly, the matrix of the environmental model in this experiment is designed as $A = \text{diag}\{-\frac{Ge}{Ce}, U\}$ and $B = \left[-\frac{1}{Ce}, 0\right]^T$ where $Ge = 7$, $Ce = -0.4$ and $U = 0.001$.

The experimental results have been presented in Figs. 8.13 and 8.14. From Fig. 8.13, it can be seen that the desired tracking performance and prescribed constraint tracking error can be satisfied with the usage of the proposed control which is consistent with simulation results in the last section. Besides, the number of neural nodes c_D, c_C and c_G are increased adaptively and remain stable at 14, 35, and 38

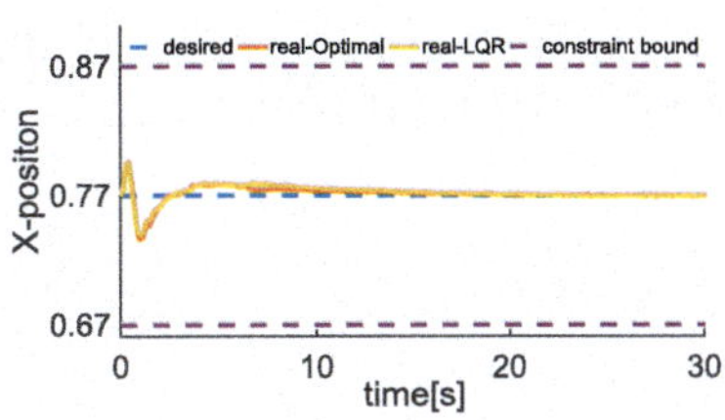

(a) Comparison of tracking results for X-axis

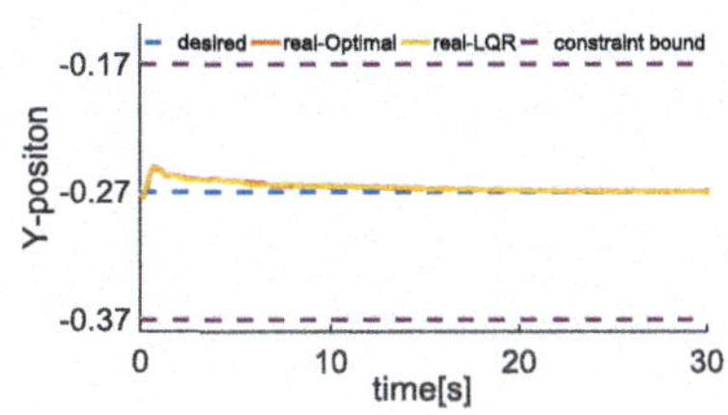

(b) Comparison of tracking results for Y-axis

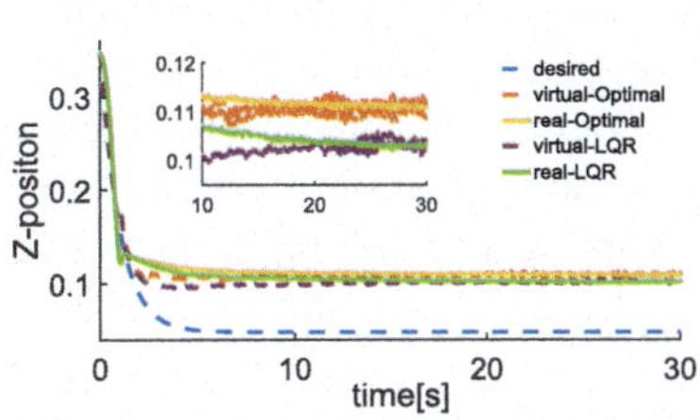

(c) Comparison of tracking results for Z-axis

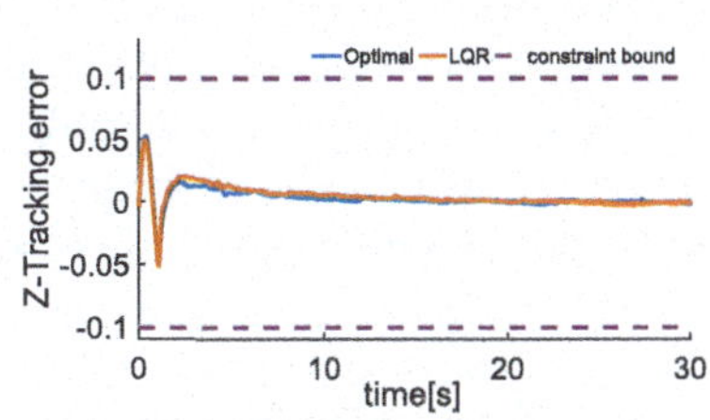

(d) Comparison of tracking error for Z-axis

Fig. 8.13 Comparison of tracking results between the improved optimal impedance control and conventional LQR method. "desired" denotes the desired trajectory which is defined in (8.48); "virtual" represents the trajectory which is adjusted by the impedance model in terms of the desired trajectory

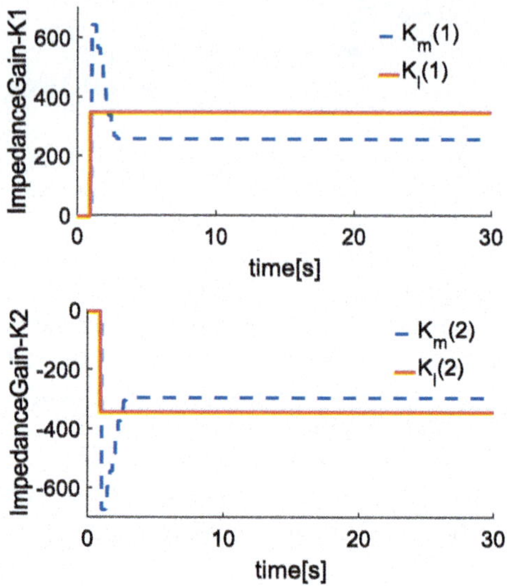

Fig. 8.14 Evolution of the experimental impedance gain K_m and K_l

eventually. In the presence of a sponge, the improved adaptive impedance control leads to more flexible interaction behaviour such that less compression of the sponge occurs as shown in Fig. 8.13c. Figure 8.14 shows the evolution result of K_m and K_l which are calculated by the developed algorithm and LQR. From these results, it seems more convenient and efficient for the user to employ the proposed method to design an FNN impedance controller since we do not need to have environmental knowledge or specify the structure of a fuzzy neural network beforehand.

8.6 Conclusion

In this chapter, we have developed a novel fuzzy neural network learning methodology based on broad learning to tackle the optimization problem of robot-environment interaction. The incremental broad fuzzy neural network, which can augment the membership and enhancement node adaptively, is employed to approximate the unknown dynamic model of the manipulator. Under the constraint Lyapunov function, the closed-loop stability of the system has been proven and the transient performance is satisfied by using the proposed controller. The impedance model of the environment can be obtained adaptively and the problem of discontinuity during the impedance optimization has been addressed by the improved optimization scheme. Simulation and experimental results have illustrated that the proposed framework is valid and efficient for incremental motor control and adaptive impedance control under the unknown environment model.

References

1. Yang C, Jiang Y, Li Z et al (2016) Neural control of bimanual robots with guaranteed global stability and motion precision. IEEE Trans Indus Inf 13(3):1162–1171
2. Liu X, Yang C, Chen Z et al (2018) Neuro-adaptive observer based control of flexible joint robot. Neurocomputing 275:73–82
3. Zhang X, Liu J, Feng J et al (2019) Effective capture of nongraspable objects for space robots using geometric cage pairs. IEEE/ASME Trans Mechatr 25(1):95–107
4. Haghighi DA, Mobayen S (2018) Design of an adaptive super-twisting decoupled terminal sliding mode control scheme for a class of fourth-order systems. ISA Trans 75:216–225
5. Mobayen S, Tchier F (2015) A new LMI-based robust finite-time sliding mode control strategy for a class of uncertain nonlinear systems. Kybernetika 51(6):1035–1048
6. Liu YJ, Li J, Tong S et al (2016) Neural network control-based adaptive learning design for nonlinear systems with full-state constraints. IEEE Trans Neural Netw Learn Syst 27(7):1562–1571
7. Huang H, Zhang T, Yang C, Chen CP (2019) Motor learning and generalization using broad learning adaptive neural control. IEEE Trans Indus Electron 67(10):8608–8617
8. Mobayen S, Majd VJ (2012) Robust tracking control method based on composite nonlinear feedback technique for linear systems with time-varying uncertain parameters and disturbances. Nonlinear Dyn 70:171–180
9. Xi X, Mobayen S, Ren H et al (2018) Robust finite-time synchronization of a class of chaotic systems via adaptive global sliding mode control. J Vib Control 24(17):3842–3854
10. Dong W, Zhao Y, Chen Y et al (2011) Tracking control for nonaffine systems: a self-organizing approximation approach. IEEE Trans Neural Netw Learn Syst 23(2):223–235
11. Wang HO, Tanaka K, Griffin MF (1996) An approach to fuzzy control of nonlinear systems: stability and design issues. IEEE Trans Fuzzy Syst 4(1):14–23
12. He W, Kong L, Dong Y et al (2017) Fuzzy tracking control for a class of uncertain MIMO nonlinear systems with state constraints. IEEE Trans Syst Man Cybern Syst 49(3):543–554
13. Wai RJ, Muthusamy R (2017) Adaptive fuzzy neural network control for a constrained robot using impedance learning. IEEE Trans Neural Netw Learn Syst 29(4):1174–1186
14. Yang C, Jiang Y, Na J et al (2018) Finite-time convergence adaptive fuzzy control for dual-arm robot with unknown kinematics and dynamics. IEEE Trans Fuzzy Syst 27(3):574–588
15. Peng H, Li F, Liu J et al (2019) A symplectic instantaneous optimal control for robot trajectory tracking with differential-algebraic equation models. IEEE Trans Indus Electron 67(5):3819–3829
16. Yang C, Jiang Y, He W et al (2018) Adaptive parameter estimation and control design for robot manipulators with finite-time convergence. IEEE Trans Indus Electron 65(10):8112–8123
17. Lai G, Liu Z, Zhang Y et al (2015) Robust adaptive fuzzy control of nonlinear systems with unknown and time-varying saturation. Asian J Control 17(3):791–805
18. Han H, Qiao J (2010) A self-organizing fuzzy neural network based on a growing-and-pruning algorithm. IEEE Trans Fuzzy Syst 18(6):1129–1143
19. Wai RJ, Muthusamy R (2012) Fuzzy-neural-network inherited sliding-mode control for robot manipulator including actuator dynamics. IEEE Trans Neural Netw Learn Syst 24(2):274–287
20. Chen CLP, Liu Z (2017) Broad learning system: an effective and efficient incremental learning system without the need for deep architecture. IEEE Trans Neural Netw Learn Syst 29(1):10–24
21. Igelnik B, Pao YH (1995) Stochastic choice of basis functions in adaptive function approximation and the functional-link net. IEEE Trans Neural Netw 6(6):1320–1329
22. Liu Z et al (2017) Broad learning system: feature extraction based on K-means clustering algorithm. Paper presented at the 2017 4th international conference on information, cybernetics and computational social systems (ICCSS), Dalian, China, 24–26 July 2017
23. Feng S, Chen CLP (2018) Fuzzy broad learning system: a novel neuro-fuzzy model for regression and classification. IEEE Trans Cybern 50(2):414–424

24. Han M, Feng S, Chen CLP et al (2018) Structured manifold broad learning system: a manifold perspective for large-scale chaotic time series analysis and prediction. IEEE Trans Knowl Data Eng 31(9):1809–1821
25. Burdet E, Osu R, Franklin DW et al (2001) The central nervous system stabilizes unstable dynamics by learning optimal impedance. Nature 414(6862):446–449
26. Franklin DW, Liaw G, Milner TE et al (2007) Endpoint stiffness of the arm is directionally tuned to instability in the environment. J Neurosci 27(29):7705–7716
27. Chib VS, Patton JL, Lynch KM et al (2006) Haptic identification of surfaces as fields of force. J Neurophysiol 95(2):1068–1077
28. Yang C, Peng G, Li Y et al (2018) Neural networks enhanced adaptive admittance control of optimized robot-environment interaction. IEEE Trans Cybern 49(7):2568–2579
29. Li Z, Huang Z, He W et al (2016) Adaptive impedance control for an upper limb robotic exoskeleton using biological signals. IEEE Trans Indus Electron 64(2):1664–1674
30. Li Z, Liu J, Huang Z et al (2017) Adaptive impedance control of human-robot cooperation using reinforcement learning. IEEE Trans Indus Electron 64(10):8013–8022
31. Johansson R, Spong MW (1994) Quadratic optimization of impedance control. Paper presented at the Proceedings of the 1994 IEEE international conference on robotics and automation, San Diego, CA, USA, 08–13 May 1994
32. Matinfar M, Hashtrudi-Zaad K (2005) Optimization-based robot compliance control: geometric and linear quadratic approaches. Int J Rob Res 24(8):645–656
33. Peng G, Yang C, He W et al (2019) Force sensorless admittance control with neural learning for robots with actuator saturation. IEEE Trans Indus Electron 67(4):3138–3148
34. Yang C, Peng G, Cheng L et al (2019) Force sensorless admittance control for teleoperation of uncertain robot manipulator using neural networks. IEEE Trans Syst Man Cybern Syst 51(5):3282–3292
35. Ge SS, Li Y, Wang C (2014) Impedance adaptation for optimal robot-environment interaction. Int J Control 87(2):249–263
36. Jin X, Xu JX (2013) Iterative learning control for output-constrained systems with both parametric and nonparametric uncertainties. Automatica 49(8):2508–2516
37. Rahman MM, Ikeura R, Mizutani K (2002) Investigation of the impedance characteristic of human arm for development of robots to cooperate with humans. JSME Int J Ser C Mech Syst Mach Elements Manuf 45(2):510–518
38. Bertsekas D (2012) Dynamic programming and optimal control. Athena Scientific, Belmont, MA, USA
39. Park J, Sandberg IW (1999) Identification of nonlinear dynamic systems using functional link artificial neural networks. IEEE Trans Syst Man Cybern 29(2):254–262
40. Ge SS, Wang C (2004) Adaptive neural control of uncertain MIMO nonlinear systems. IEEE Trans Neural Netw 15(3):674–692
41. Corke P (2017) Robot manipulator capability in MATLAB: a tutorial on using the robotics system toolbox [tutorial]. IEEE Rob Autom Mag 24(3):165–166
42. Shen T (2000) Robust control of robot. Tsinghua University Press, Beijing, China

The manufacturer's authorised representative in the EU is Springer Nature Customer Service Centre GmbH, Europaplatz 3, 69115 Heidelberg, Germany. If you have any concerns regarding our products, please contact ProductSafety@springernature.com

Printed and bound by CPI Group (UK) Ltd, Croydon, CR0 4YY
30/09/2026
02225265-0002